Storage, Processing, and Nutritional Quality of Fruits and Vegetables

Editor

D. K. Salunkhe

Department of Nutrition and Food Science
Utah State University
Logan, Utah

Co-editors

J. Y. Do
M. T.
H. R.
S. K.
G. G.

Published by

CRC PRESS, Inc.
18901 Cranwood Parkway · Cleveland, Ohio 44128

Library of Congress Cataloging in Publication Data

Salunkhe, D. K.
    Storage, processing, and nutritional quality of
fruits and vegetables.

    Includes bibliographies.
    1. Fruit. 2. Vegetables. 3. Fruit — Storage.
4. Vegetables — Storage. 5. Fruit processing.
6. Vegetable processing. I. Title.
TX557.S24   641.4   74-20660
ISBN-0-8493-0123-8

© 1974 by CRC Press, Inc.

Second Printing, January 1976
Third Printing, December 1976
CRC Press, Inc.

International Standard Book Number 0-8493-0123-8
Former International Standard Book Number 0-87819-125-9

Library of Congress Card Number 74-20660
Printed in the United States

AUTHOR

D. K. Salunkhe is professor of nutrition and food science, Utah State University. He was born in India in 1925. He graduated with a B.Sc. (Agri.) degree with honors from Poona University in 1949; in 1951 and 1954 he received his M.S. and Ph.D. degrees from Michigan State University. Professor Salunkhe came to Utah State University in 1953 as a research associate in the horticulture department. He has taught courses in bacteriology and plant pathology, post-harvest physiology and biochemistry of fruits and vegetables, processing of fruits and vegetables, food toxicology, and world food problems. He has authored over 200 scientific papers and reports. Some of his articles have received recognition and awards as outstanding articles in biological journals. He was chairman and speaker on many national and international scientific symposia. Professor Salunkhe sat on committees for the American Society for Horticultural Science and Institute of Food Technology whose purpose it was to select articles and research contributors worthy of awards.

He has reviewed several books for scientific journals and contributed chapters to several others. He is author of *Nutritional Quality of Fruits and Vegetables* and co-author of *Post-harvest Physiology and Biochemistry of Fruits and Vegetables* which will be published early in 1975.

Professor Salunkhe has traveled extensively around the world to visit prominent research laboratories and to deliver lectures on research conducted in his laboratories at Utah State University. He was an Alexander Humboldt Senior Fellow and a guest professor at the Technical University in Karlsruhe, West Germany. He was a guest lecturer at the Technological Institute, Moscow, USSR, and an exchange scientist to Czechoslovakia on behalf of National Academy of Science. He was on the advisory committee for the International Congress of Food Science and Technology, National Academy of Sciences, National Research Council, and several food processing and consumer organizations.

This book originally appeared as articles in Volume 4 Issues 1, 2, and 4 and Volume 5 Issue 1 of *CRC Critical Reviews in Food Technology,* a quarterly journal published by CRC Press, Inc. We would like to acknowledge the editorial assistance received by the journal's editor, Mr. Thomas Furia, Dynapol Palo Alto, California. The referees for these articles were Willard B. Robinson, Cornell University, Geneva, New York; M. W. Miller, University of California, Davis; S. K. Ries, Michigan State University, East Lansing; and A. R. Rahman, U.S. Army Natick Laboratories, Natick, Massachusetts.

PREFACE

The main purpose of this book is to review and bring into focus the nutritional value and quality of fruits and vegetables as influenced by chemical treatments, storage, and processing conditions. Fruits and vegetables play an important role by providing nutritionally essential ingredients in the human diet. They are sources of minerals, vitamins, carbohydrates, proteins, and fats, and also of significant value as roughage which promotes digestion and helps to prevent constipation. Some of the fruits and vegetables are important in neutralizing the acids produced in the course of digestion of meat and dairy products.

Per capita consumption of fruits and vegetables has increased since 1900. The continued increase in population should create a market for more fruits and vegetables than have been presently produced. Consumer food buying patterns have rapidly changed in recent years. Now, over 30% of the food consumed by Americans consists of fruits and vegetables. The consumption of tomatoes and lettuce has doubled in the last 25 years.

As the U.S. population continues to urbanize, the market for fresh commodities is increasingly removed from rural production centers and transportation plays a much greater part in the overall scheme. Air freight shipment even allows certain fruits such as strawberries and vegetables such as lettuce to be raised in the U.S. and sold in Europe, Australia, and Japan. Hence growers, shippers, wholesalers, and retailers have to prolong the storage life of fresh produce and maintain its high quality without pricing themselves out of business.

The "Green Revolution" has increased the yield of food plants. The quality of fruits and vegetables could be improved by genetical approaches and by chemical treatments during their production in the field. Thus, we could have nutritionally better-quality fresh produce to start with on consumption, storage, and processing.

Losses due to postharvest deterioration and diseases of fresh produce affect over 40% of the produce harvested in the world that is never consumed because of spoilage. In underdeveloped countries where modern facilities of storage, prepackaging, and processing are less common, this figure could be over 50%. In addition, consumption of processed (frozen, canned, fermented, and dehydrated) fruits and vegetables has significantly increased in recent years.

The urgent demand for enormous quantities of processed fruits and vegetables after World War II was the important driving force for progress in the technology of processing and preservation of food. Consequently, improved technological processes and equipment and advances in the scientific knowledge pertaining to processing methods and their effects on quality and nutritional value were developed and became available to the producers as well as processors.

The recent renewed interest of the general public in the nutritional quality of food has prompted us to review the nutritional value and quality of fruits and vegetables in fresh, as well as processed, conditions. Various methods are being employed to prolong the storage life of fruits, vegetables, and their products for both short and long terms, such as low temperature, precooling, waxing, packaging, ionizing radiation, chemical treatments, modified atmosphere, controlled atmosphere, subatmosphere, freezing, fermentation, canning, and dehydration. The intelligent application of these scientific methods in the fruit and vegetable products is highly essential. Although notable advances have been made in the knowledge of nutrition and quality of fruits and vegetables, there remains a vast amount of research to be done to deliver excellent quality and high nutritional value of fruits and vegetables and their products to the consumer.

D. K. Salunkhe
Logan, Utah

TABLE OF CONTENTS

PART I. ASSESSMENT OF NUTRITIVE VALUE, QUALITY, AND STABILITY OF CRUCIFEROUS VEGETABLES DURING STORAGE AND SUBSEQUENT TO PROCESSING

PART II. DEVELOPMENTS IN TECHNOLOGY AND NUTRITIVE VALUE OF DEHYDRATED FRUITS, VEGETABLES, AND THEIR PRODUCTS

PART III. THE USE OF CERTAIN CHEMICALS TO INCREASE NUTRITIONAL VALUE
AND TO EXTEND QUALITY IN ECONOMIC PLANTS

PART IV. DEVELOPMENTS IN TECHNOLOGY OF STORAGE AND HANDLING OF
FRESH FRUITS AND VEGETABLES

ASSESSMENT OF NUTRITIVE VALUE, QUALITY, AND STABILITY OF CRUCIFEROUS VEGETABLES DURING STORAGE AND SUBSEQUENT TO PROCESSING

Authors: **D. K. Salunkhe**
            S. K. Pao
            Department of Nutrition and Food Science
            Utah State University
            Logan, Utah

            G. G. Dull
            Richard B. Russell Agricultural Research Center
            Agricultural Research Service
            United States Department of Agriculture
            Athens, Georgia

INTRODUCTION

The cruciferae family provides a notable group of vegetables such as cabbage (*Brassica oleracea* L. var. *capitata* L.), cauliflower (*B. oleracea* L. var. *botrytis* L.), kale and collard (*B. oleracea* L. var. *acephala* D.C.), Brussels sprout (*B. oleracea* L. var. *gemmifera* Zenker), sprouting broccoli (*B. oleracea* L. var. *italica* Plenk), kohlrabi (*B. caulorapa* Pasq.), turnip (*B. rapa* L.), rutabaga (*B. napobrassica* Mill.), Chinese cabbage, pe-tsai (*B. pekinensis* Rupr.), pak-choi (*B. chinensis* L.), mustard (*B. juncea* Coss; *B. hirta* Moench), watercress (*Nasturtium officinale* R. Br.), horseradish (*Armoracia lapathifolia* Gilib), and radish (*Raphanus sativus* L.). The flowers of these plants have four sepals and four petals, hence the name cruciferae, in reference to the similarity to the Greek cross.

It is believed that modern cabbage originated from wild types that grow on the cliffs of eastern England and along the coast of Denmark and northwestern France and various other localities such as Greece, Italy, Egypt, and the Mediterranean and Asia Minor regions. Wild cabbage which is similar to collard shows variations in general appearance and foliage and is still used for human food. Broccoli and head cabbage have developed from these wild types while the Brussels sprout is a cabbage mutant.

Most cruciferous vegetable crops are hardy and grow profusely in cool weather. They grow adequately in all types of soil, but fertile, heavy soils having high-moisture retention are the most desirable. Since the crops of this family are closely related, their cultural requirements and susceptibility to several diseases and insects are similar. They need a moderate amount of fertilizer and usually respond to nitrogen. The amounts of

1

nitrogen, phosphorus, and potassium fertilizer depend upon the environmental factors and soil composition. Copper, manganese, zinc, and boron are essential trace elements required to yield high-nutritional value and high-quality fresh produce.

Some of these crops are important in home gardens; therefore, they have a place in urbanized population centers. Likewise, some of them appear to be beautiful and colorful plants which have ornamental value in addition to nutritional benefits.

The world is faced with a need for all possible production of highly nutritive food. New directions and increased consciousness in NUTRITION will have a significant impact upon modern agriculture and food industry. Now is an appropriate time to assess production, storage, and processing and to determine whether or not crops with more nutrients can be produced and preserved for a longer time without loss in quality and wholesomeness.

The main purpose of this paper is to assess the nutritive ability of cruciferous vegetables, both fresh and processed, to fulfill the consumer's needs for vitamins, minerals, proteins, and calories. On the production side are problems of crop adaptability, growing time, productivity, labor, and equipment; on the consumption side are palatability, toxicity, nutritive value, cost and regular supply. The last point involves storage and processing. This comparison of the crops should be viewed as tentative and partial rather than final and complete.

The data presented here may assist in future decisions as to the relative importance of these vegetables when compared with other commodities. The fruits and vegetables are evaluated according to their composition and to the dietary needs of man. Data have been compiled on the levels of energy (calories), fat, carbohydrates, proteins, calcium, phosphorus, potassium, magnesium, sodium, iron, vitamin A, riboflavin, niacin, thiamine, ascorbic acid, and wherever available on other vitamins, amino acids, biological value, and digestibility.

NUTRIENTS AND QUALITY
OF CRUCIFEROUS VEGETABLES
AND THEIR PRODUCTS

As the vegetables undergo transportion and storage or processing enroute to the consumer's table, they suffer losses in quantity, quality, and, particularly, in nutritive values which are frequently destroyed or denatured. The specific effects on nutrients are dependent upon certain factors such as oxygen; light; acidity; time; temperature; enzymes; population, genus, and species of microorganisms; presence of metals; structure and texture of product; handling; container; processing; storage; and interactions of these.

Horticultural Considerations

The nutrient contents of fresh vegetables usually depend upon factors such as cultivars, environment, and maturity and are synthesized in a plant or in its parts while growing on a farm.

Cultivars

In the past, geneticists and plant breeders were concerned mainly with breeding new cultivars primarily to achieve increased yield and disease resistance and/or to enhance color, flavor, and texture. In most breeding programs, nutritive value has often been merely an afterthought. However, this attitude is now changing. The recent withdrawal of the "Lenape" cultivar of potatoes by the U.S. and Canadian Departments of Agriculture due to its unusually high concentrations of solanine, and the Food and Drug Administration now requiring that new cultivars meet nutritional and toxicant requirements in order to be classified "GRAS" (Generally Recognized As Safe) means that plant breeders must become more actively concerned with the nutritive composition of their breeding lines. In recent years, factors such as high protein in wheat, vitamin C and carotene in tomatoes, and lysine and tryptophan in corn have been taken into breeding consideration (Stevens, 1973).

Burrell et al. (1940) found a 3.5-fold range in the ascorbic acid content of 31 cultivars of cabbage. Walker and Foster (1946) concluded that selection and genetic manipulation could increase the ascorbic acid content of vegetables considerably.

Environment

Some of the ecological, cultural, and physical factors significantly influence the chemical and nutritional composition and anatomical and morphological structure of plants.

There is a mechanism in plants which can convert sucrose and hexoses into ascorbic acid. Light, temperature, and carbon dioxide, however, influence this mechanism and eventual accumulation of the ascorbic acid. Precursor(s) of ascorbic acid are produced during the process of photosynthesis.

Light

Hamner and Parks (1944) found 28.2 mg ascorbic acid in 100 g of turnip greens after a 200-foot-candle light exposure and 235 mg per 100 g after exposure to 5,000 foot candle. Reder et al. (1943b) studied two varieties of turnip greens and found 191 mg ascorbic acid per 100-g sample when plants were grown in the sunshine during 49% of the growing season and only 128 mg in those grown in a smaller amount of sunshine. Somers and Kelly (1957) reported that turnips grown in full sunlight contained more ascorbic acid and less carotene than those grown in the shade. Turnips harvested in the morning have been reported to contain more riboflavin than those harvested at other times of the day. Thiamine synthesis in plants is stimulated by light and generally occurs in the leaves, and as a rule, increases in concentration until the plant is mature. Light stimulates plant growth which in turn increases the demand for iron. In general, the effects of light on the synthesis of any nutrient in a plant depend upon plant species, amount and duration of light, temperature, and soil fertility.

Temperature

The optimum temperature for rapid growth by a species of edible plant does not always equate with its optimum for synthesizing and accumulating nutrients. Furthermore, the specific temperature promoting the greatest translocation, synthesis, and accumulation of one nutrient is often different for another. In general, the content of ascorbic acid in turnip greens held in the dark was decreased as the temperature was increased from 50 to 86°F (Somers et al., 1948). Gustafson (1950) noted a higher riboflavin content in broccoli and cabbage when grown at 50 to 59°F and they also had the highest thiamine content at that same temperature. Gustafson noted species differences in thiamine content at these temperatures.

Season and Climate

Pooles et al. (1944) and Smith et al. (1948) reported increases in the quantity of ascorbic acid in cabbage as the growing season progressed. On the other hand, according to Pyke (1942), cabbage harvested early in the season had five times more vitamin C than cabbage harvested late. Hansen (1945a) reported that climatic changes did not affect the ascorbic acid values of broccoli, collard, and kale. However, he found that the carotene content of carrots tended to decrease during fall and winter and increase in the early spring (1945b).

As the literature concerning season and climate is limited and conflicting, it is difficult to draw a definite conclusion as to the relation of season to the nutrient content.

Location, Soil Fertility, and Irrigation

The ascorbic acid contents of turnips (Reder et al., 1943a), cabbage (Janes, 1944), and rutabaga (Nylund, 1949) were influenced by the locations of the crops which no doubt were influenced by variations in soil fertility and moisture. Aberg and Ekdahl (1948) found less ascorbic acid in kale with suboptimal nitrogen than when it was adequately fertilized. Burrel et al. (1940), working with cabbage and Sheets et al. (1954), with collard, observed that ascorbic acid content increased with nitrogen fertilization. Sheets et al. (1954), however, found that the carotene in collard increased because of nitrogen-deficient soil. In general, it is evident that nitrogen fertilization of soil markedly increases the crop yield but has only a mild or insignificant effect upon its nutrient content.

The effects of calcium fertilization depend upon the pH of the soil. Hansen (1945a) noted no effects from a limestone treatment on the calcium content of kale, whereas Maynard et al. (1964) noted a marked increase. Sheets et al. (1944) conducted an extensive study of liming and observed a slight increase in calcium in turnips but found no significant differences in the calcium content in collard following liming (Sheets et al, 1954). Plants grown with a high concentration of sulfate in the soil contained higher quantities of both inorganic and organic sulfur than those grown with low concentrations.

Eheart (1966) noted that the application of nitrogen at the rate of 100 lb per acre decreased chlorophyll a, total chlorophyll, and carotene. Eheart et al. (1955) had previously shown that the ascorbic acid content of turnip greens was highest in the spring and lowest in the fall; carotene content decreased with high light intensity; thiamine content increased when trace elements were included in the fertilizer; and that riboflavin content, which also increased with high light intensity, evidenced an additional increase when the light was combined with an application of trace elements to the soil. Holmes et al. (1945) reported that the addition of 150 lb of magnesium sulfate per acre significantly increased the magnesium content of the kale and slightly increased its calcium and phosphorus. Limestone applied at the rate of 1,000 lb per acre increased the calcium and phosphorus content of the kale but decreased its iron content. Neither magnesium sulfate nor limestone had any significant effects on the carotene content.

Del Valle and Harmon (1967) showed that collard growth was influenced by soil temperature because of the increased uptake of minerals, and the phosphorus and potassium contents were higher in leaves when large quantities of minerals were available in the soil. They further indicated that the calcium and phosphorus contents of collard were increased by soil temperatures of 75 and 85°F and decreased at 95°F.

Sheets et al. (1955) stated that the phosphorus and carotene content increased in turnip greens with irrigation.

The literature relating to the effects of fertilization upon the nutrient contents is voluminous and generally contradictory, probably because of variations in soil chemistry, water availability, and other environmental factors. It can be stated, however, that adequate fertilization with nitrogen, phosphorus, potassium, calcium, magnesium, manganese, boron, and iron is essential for normal plant growth and adequate yields of high-quality nutritious crops.

Maturity

Small cabbage heads are higher in ascorbic acid than large ones (Branion et al., 1948). Speirs et al. (1951) noted that small turnip leaves had relatively more thiamine and riboflavin content than ascorbic acid but less carotene than large leaves. In general, broccoli, cabbage, and cauliflower had the maximum content of nutritive value at harvesttime.

It is exceedingly important that vegetables in general and crucifers in particular be harvested at the proper time. The exact degree of maturity at which a given cultivar or species should be harvested depends upon the purpose for which it is used. When cabbage or a cauliflower head is harvested for the local fresh market, it must be harvested at the prime maturity. When it is harvested for shipping or processing, it should be harvested a week early. If it is harvested too early and then stored for a length of time, it may shrivel and the characteristic aroma will disappear; and if harvested too late and then stored for a longer period, it might have higher than average concentrations of sulfur compounds and less ascorbic acid and other water-soluble vitamins. Explicit directions and measurements cannot be provided as to the best time to harvest crucifers. No one index of maturation is infallible, and often a combination of several indices and, above all, experience are essential.

Production (Tonnage and Nutrients) Considerations

Horticultural commodities such as fruits and vegetables are generally bought and consumed primarily for flavor and appearance rather than for nutritive value. This concept must be changed by a cross between "Nutrition" and "Food Science" and by educating the public to the "hybrid vigor" produced by such a cross breeding. For their health and welfare, consumers must become aware of the importance of fruits and vegetables in general and cruciferous vegetables in particular (Table 1). This publication contains comparative basic information pertaining to the nutritive contribution of 12 major fruits, 12 cruciferous vegetables, and 18 other major vegetables (Table 2).

The extent to which a food contributes nutrients to man's total food supply is governed both by the *amount of nutrients* and the *quantity of the food* that is consumed.

These production statistics establish that the cruciferous vegetables are in poorly consumed category. Foods that are produced and consumed in large quantities such as potatoes, tomatoes, and oranges can significantly contribute to the well-being of a nation by supplying calories and a certain amount of some nutrients even though their relative nutritive value is not high. Despite the fact that among 42 fruits and vegetables surveyed for this study potatoes rank first in annual production of riboflavin, a person would have to eat about 7.7 lb per day to meet the minimum daily dietary allowance for an adult. In

TABLE 1

Nutrients in Cruciferous Vegetables and Their Major Functions in the Human Body

Nutrients	Functions
Proteins	Contain essential amino acids — isoleucine, leucine, lysine, methionine, phenylalanine, threonine, tryptophan, and valine; provide nitrogen for hormones, enzymes, blood, and calories; form network structure in bones; act as a biological buffer system; have immunological effects.
Carbohydrates	Source of energy; provide roughage (cellulose and hemicellulose) to aid elimination.
Fats	Supply energy; are carriers for certain vitamins; contain linoleic acid, an essential fatty acid, which acts as a precursor for several other fatty acids; required for growth and dermal integrity; when deposited, function as protective tissue.
Vitamins	
A	Maintains healthy mucous membranes (eyes and mouth and gastrointestinal, respiratory, and genito-urinary systems); protects against infection; incorporated in rhodopsin (eye pigment); essential for light/dark adaptation of vision; promotes normal skin and tooth growth; maintains healthy gums and teeth; stimulates reproduction and lactation; is involved in bone protein formation.
B_1 (thiamine)	Essential for normal activities of the nervous system; coenzyme in carbohydrate metabolism.
B_2 (riboflavin)	Necessary for healthy eyes; combined with protein, accelerates recovery from injury and infection; coenzyme in oxidative metabolism.
Niacin	Supplies energy during injury, fever, vomiting, diarrhea, intestinal disorders, and pregnancy; aids in proper maintenance of skin; activates energy sources in many biological reactions.
B_6 (pyridoxine)	Aids in formation of red blood cells; converts tryptophan to serotonin and niacin; prevents nerve inflammation during drug treatment; is involved in amino acid and lipid metabolism and utilization of proteins.
Pantothenic acid	Necessary for synthesis of sterols required as components for central nervous system; aids in the utilization of riboflavin and pyridoxine in tissue repair; essential for growth and skin; present in red blood cells and plasma; involved in the synthesis of porphyrin; is the essential part of coenzyme A thus participating in all important metabolic activities.
Folacin (folic acid group)	Along with iron, vitamins C and B_{12}, is necessary for hematopoisesis (blood formation); has therapeutic value for anemia in infancy and pregnancy; hastens maturation of bone marrow.
C	Increases resistance to infection; essential for normal development and maintenance of collagen; aids in absorption of iron and maintains connective tissues in blood capillaries, thus hastens the process of healing wounds and prevents small blood vessel hemorrhages; essential for formation of bone protein.
E	Protects vitamin A and many other potential oxidants from oxidation thus maintaining their stability.
K	Required for the formation of prothrombin (blood clot) and for normal function of liver.
Minerals	
Calcium	Regulates nerve and muscle activity; strengthens bones and teeth; essential for blood coagulation; needed to maintain proper permeability of cell membranes; facilitates phosphorus absorption.
Phosphorus	Component of complex fat in nerve tissue and in neural metabolic process; has many functions similar to that of calcium.
Potassium	Located mainly in the intercellular fluids; acts with sodium to provide proper nerve stimulation and to preserve the optimum cyto-osmosis and permeability of the cell membranes.
Sodium	Located chiefly in the intercellular fluids; interrelated with potassium.
Manganese	Regulates utilization of thiamine.
Iron	Constituent of blood and several enzymes; controls anemia; important in the respiratory electron transport.
Magnesium	Required for nerve and muscle response, along with calcium; connected with calcium metabolism; cofactor in many enzymatic reactions.
Copper	Aids in bone formation and the absorption of iron from alimentary tract; needed for the iron-containing enzymes and hemoglobin formation.
Cobalt	Essential for synthesis of vitamin B_{12} and for red blood cell formation.

TABLE 1 (Continued)

Nutrients in Cruciferous Vegetables and Their Major Functions in the Human Body

Nutrients	Functions
Sulfur	Necessary for growth and replacement of skin, hair, and nails and for the formation of sulfur-containing amino acids.
Zinc	Implicated in tissue repair and skeletal growth; is a part of the chemical structure of insulin; regulates copper utilization.
Fluorine	Required for maintenance of bone structure and control of tooth decay.

Main source: Recommended Dietary Allowances, National Academy of Sciences, 1968,
Washington, D.C.

TABLE 2

Nutritional Survey of the Following 42 Fruits and Vegetables

Fruits	Cruciferous vegetables	Other vegetables	
Apricot	Cabbage	Carrot	Cantaloupe
Peach	Cauliflower	Sweet potato	Spinach
Orange	Broccoli	Tomato	Pea
Grapefruit	Brussels sprout	Sweet corn	Asparagus
Plum	Kale	Pepper	Snap bean
Grape	Watercress	Lettuce	Beet
Sour cherry	Mustard green	Potato	
Apple	Turnip green	Squash	
Strawberry	Collard	Onion	
Watermelon	Rutabaga	Cucumber	
Pear	Chinese cabbage	Lima bean	
Banana	Kohlrabi	Celery	

1970, the production of potatoes, the most prevalent horticultural crop, was about 16.5 million tons; that of oranges, 8 million tons; and tomatoes, 6 million tons; while fewer than 1.5 million tons of cabbage were harvested. Cabbage is the only cruciferous vegetable listed among the leading 15 fruits and vegetables (Figure 1). In the last 25 years, however, cabbage acreage has progressively declined; consequently, the per capita consumption in 1970 was less than half that in 1944 (USDA, Agricultural Statistics, 1972).

Taking both of these factors into consideration, calculations have been made on the proportions of certain nutrients in the food supply that are attributable to fruits and vegetables. Such information for the year 1971, obtained from National Food Situation 138, November 1971, is summarized in Figure 2, which shows total and relative percentages of the nutrients in the food supply that are derived from fruits and vegetables. Over 90% of ascorbic acid, 50% of vitamin A, 35% of riboflavin, 25% of magnesium, 20% of thiamine and niacin, and 20% of iron are provided by fruits and vegetables. Fruits and vegetables also contribute less than 1% of fat, 10% of food energy, and 7% of the protein of the total food intake.

Per capita consumption of fruits and vegetables in 1971 was about 530 lb and over half of this amount was consumed in the processed form.

The United States Department of Agriculture Handbook No. 8 (1963), Home Economic Research Report No. 36 (1969), FAO: Nutritional Studies No. 24, and other pertinent references were used to assess the nutritive value of 42 fruits and vegetables. Each commodity was then ranked for each nutrient from 1 equaling the highest to 42 the lowest amount of nutrients. These numbers are presented in parentheses in Table 3.

Energy: Fats and Carbohydrates

The body requires food energy for basal metabolism, synthesis of body tissues, physical activity, excretory processes, and to maintain thermal, physiological, and psychological balance.

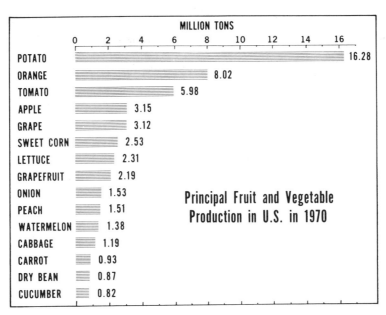

FIGURE 1. Yield of the 15 most prevalent fruits and vegetables in the United States of America in 1970.

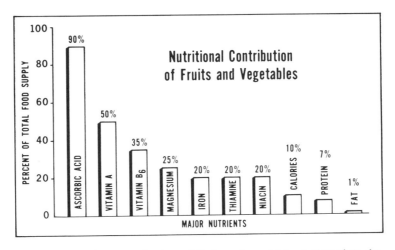

FIGURE 2. Nutrient contribution of fruits and vegetables compared to the percent of total food supply in 1971.

Caloric allowances are established to provide energy and to maintain body weight and health. The caloric adjustments are made according to the size, age, and activity of a person, to special demands during illness, pregnancy, and lactation, and to the climate where one lives.

Fats and carbohydrates are the most important sources of food energy. According to the Food and Nutrition Board, National Research Council (1968), the American diet contains about 41% fats and 47% carbohydrates of the total calories.

It is apparent from Table 4 that most of the cruciferous vegetables are low in calorics, fat, and carbohydrates. The contribution to calories is primarily due to fat and fatty acids and the predominant fatty acids in cruciferous vegetables are linoleic and linolenic. These fatty acids are essential for the human body.

Proteins: Quantity, Quality, and Utilization

The question of how much protein should be provided at the different stages of a person's

TABLE 3

Composition and Ranking of 42 Fresh Fruits and Vegetables per 100 g Edible Portion.

No.	Food	Water (%)	Energy (cal)	Protein (%)	Fat (%)	Carbohydrate (%)	Calcium (mg)	Phosphorus (mg)	Iron (mg)
						Fruits			
1.	Apricot	85.3	51(13)	1.0(32)	.2(28)	12.8(12)	17(29)	23(30)	.5(30)
2.	Peach	89.1	38(21)	.6(38)	.1(38)	9.7(17)	9(36)	19(34)	.5(29)
3.	Orange	86.0	49(14)	1.0(31)	.2(27)	12.2(13)	41(14)	20(33)	.4(37)
4.	Grapefruit	88.4	41(20)	.5(40)	.1(39)	10.6(15)	16(30)	16(38)	.4(35)
5.	Plum	81.1	66(8)	.5(41)	.2[a](29)	17.8(5)	18(28)	17(36)	.5(31)
6.	Grape	81.6	69(7)	1.3(21)	1.0(1)	15.7(7)	16(32)	12(39)	.4(39)
7.	Sour Cherry	83.7	58(11)	1.2(25)	.3(17)	14.3(11)	22(26)	19(35)	.4(38)
8.	Apple	84.4	58(10)	.2(42)	.6(5)	14.5(9)	7(40)	10(42)	.3(41)
9.	Strawberry	89.9	37(23)	.7(36)	.5(6)	8.4(21)	21(27)	21(32)	1.0(15)
10.	Watermelon	92.6	26(33)	.5(39)	.2(21)	6.4(27)	7(39)	10(41)	.5(26)
11.	Pear	83.2	61(9)	.7(37)	.4(10)	15.3(8)	8(37)	11(40)	.3(42)
12.	Banana	75.7	85(4)	1.1(30)	.2(30)	22.2(2)	8(38)	26(29)	.7(20)
						Vegetables			
13.	Carrot	88.2	42(19)	1.1(28)	.2(26)	9.7(18)	37(16)	36(21)	.7(18)
14.	Sweet Potato	70.6	114(2)	1.7(17)	.4(12)	26.3(1)	32(19)	47(16)	.7(22)
15.	Tomato	93.5	22(36)	1.1(27)	.2(19)	4.7(35)	13(34)	27(27)	.5(25)
16.	Sweet Corn	72.7	96(3)	3.5(7)	1.0(2)	22.1(3)	3(42)	111(3)	.7(21)
17.	Pepper	93.4	22(35)	1.2(24)	.2(20)	4.8(34)	9(35)	22(31)	.7(17)
18.	Lettuce	95.1	14(41)	1.2(22)	.2(18)	2.5(42)	35(18)	26(28)	2.0(5)
19.	Potato	79.8	76(6)	2.1(14)	.1(42)	17.1(6)	7(41)	53(12)	.6(24)
20.	Squash	94.0	19(38)	1.1(26)	.1(34)	4.2(37)	28(20)	29(23)	.4(32)
21.	Onion	89.1	38(22)	1.5(19)	.1(37)	8.7(20)	27(21)	36(20)	.5(28)
22.	Cucumber	95.1	15(40)	.9(33)	.1(31)	3.4(39)	25(23)	27(26)	1.1(11)
23.	Spinach	90.7	26(32)	3.2(8)	.3(14)	4.3(36)	93(7)	51(14)	3.1(1)
24.	Lima Bean	67.5	123(1)	8.4(1)	.5(8)	22.1(4)	52(10)	142(1)	2.8(3)
25.	Pea	78.0	84(5)	6.3(2)	.4(11)	14.4(10)	26(22)	116(2)	1.9(6)
26.	Asparagus	91.7	26(31)	2.5(12)	.2(23)	5.0(32)	22(25)	62(8)	1.0(14)
27.	Cantaloupe	91.2	30(27)	.7(35)	.1(35)	7.5(23)	14(33)	16(37)	.4(34)
28.	Snap Bean	90.1	32(24)	1.9(16)	.2(24)	7.1(25)	56(9)	44(17)	.8(16)
29.	Beet	87.3	43(18)	1.6(18)	.1(40)	9.9(16)	16(31)	33(22)	.7(19)
30.	Celery	94.1	17(39)	.9(34)	.1(33)	3.9(38)	39(15)	28(25)	.3(40)
						Crucifers			
31.	Cabbage	92.4	24(34)	1.3(20)	.2(22)	5.4(30)	49(11)	29(24)	.4(33)
32.	Broccoli	89.1	32(25)	3.6(6)	.3(16)	5.9(28)	103(6)	78(7)	1.1(13)
33.	Brussels Sprout	85.2	45(17)	4.9(4)	.4(9)	8.3(22)	36(17)	80(6)	1.5(10)
34.	Cauliflower	91.0	27(30)	2.7(11)	.2(25)	5.2(31)	25(24)	56(10)	1.1(12)
35.	Kale	82.7	53(12)	6.0(3)	.8(4)	9.0(19)	249(2)	93(4)	2.7(4)
36.	Watercress	93.3	19(37)	2.2(13)	.3(13)	3.0(41)	151(5)	54(11)	1.7(8)
37.	Mustard green	89.5	31(26)	3.0(10)	.5(7)	5.6(29)	183(4)	50(15)	3.0(2)
38.	Turnip green	90.3	28(29)	3.0(9)	.3(15)	5.0(33)	246(3)	58(9)	1.8(7)
39.	Chinese Cabbage	95.0	14(42)	1.2(23)	.1(32)	3.0(40)	43(12)	40(18)	.6(23)
40.	Rutabaga	87.0	46(15)	1.1(29)	.1(41)	11.0(14)	66(8)	39(19)	.4(36)
41.	Collard	85.3	45(16)	4.8(5)	.8(3)	7.5(24)	250(1)	82(5)	1.5(9)
42.	Kolhrabi	90.3	29(28)	2.0(15)	.1(36)	6.6(26)	41(13)	51(13)	.5(27)

TABLE 3 (Continued)

Composition and Ranking of 42 Fresh Fruits and Vegetables per 100 g Edible Portion.

No.	Food	Sodium (mg)	Potassium (mg)	Magnesium (mg)	Vitamin A (IU)	Thiamine (mg)	Riboflavin (mg)	Niacin (mg)	Vitamin C (mg)
					Fruits				
1.	Apricot	1(40)	281(18)	12(34)	2,700(10)	.03(39)	.04(33)	.6(24)	10(31)
2.	Peach	1(37)	202(32)	10(38)	1,330(12)	.02(41)	.05(24)	1.0(10)	7(38)
3.	Orange	1(39)	200(33)	11(37)	200(29)	.10(12)	.04(32)	.4(30)	50(13)
4.	Grapefruit	1(38)	135(39)	12(33)	80(35)	.04(32)	.02(41)	.2(40)	38(16)
5.	Plum	2(31)	299(15)	9(39)	300(25)	.08(19)	.03(40)	.5(28)	5[a](39)
6.	Grape	3(26)	158(37)	13(30)	100(33)	.05(29)	.03(39)	.3(36)	4(42)
7.	Sour Cherry	2(29)	191(34)	14(28)	1,000(13)	.05(28)	.06(20)	.4(31)	10(32)
8.	Apple	1(41)	110(41)	8(41)	90(34)	.03(40)	.02(42)	.1(41)	4(40)
9.	Strawberry	1(36)	164(35)	12(31)	60(37)	.03(36)	.07(17)	.6(22)	59(11)
10.	Watermelon	1(35)	100(42)	8(40)	590(19)	.03(35)	.03(37)	.2(38)	7(37)
11.	Pear	2(30)	130(40)	7(42)	20(41)	.02(42)	.04(34)	.1(42)	4(41)
12.	Banana	1(42)	370(10)	33(10)	190(30)	.05(30)	.06(21)	.7(18)	10(33)
					Vegetables				
13.	Carrot	47(6)	341(12)	23(18)	11,000(1)	.06(24)	.05(25)	.6(23)	8(36)
14.	Sweet Potato	10(19)	243(27)	31(12)	8,800(4)	.10(15)	.06(22)	.6(25)	21(24)
15.	Tomato	3(25)	244(25)	14(27)	900(15)	.06(22)	.04(29)	.7(16)	23(22)
16.	Sweet Corn	10[b](18)	280(19)	48(5)	400(24)	.15(7)	.12(11)	1.7(4)	12(27)
17.	Pepper	13(14)	213(30)	18(22)	420(22)	.08(16)	.08(16)	.5(26)	128(4)
18.	Lettuce	9(20)	264(21)	11(36)	900(14)	.06(21)	.06(19)	.3(32)	8(35)
19.	Potato	3(27)	407(4)	34(9)	20[a](42)	.10(14)	.04(35)	1.5(6)	20(25)
20.	Squash	1(34)	202(31)	17(23)	410(23)	.05(26)	.09(15)	1.0(9)	22(23)
21.	Onion	10(17)	157(38)	12(32)	40(38)	.03(37)	.04(31)	.2(39)	10(29)
22.	Cucumber	6(23)	160(36)	11(35)	250(27)	.03(33)	.04(27)	.2(37)	11(28)
23.	Spinach	71(3)	470(2)	88(1)	8,100(5)	.10(10)	.20(7)	.6(21)	51(12)
24.	Lima Bean	2(33)	650(1)	67(2)	290(26)	.24(2)	.12(12)	1.4(7)	29(19)
25.	Pea	2(32)	316(14)	35(8)	640(17)	.35(1)	.14(10)	2.9(1)	27(20)
26.	Asparagus	2(28)	278(20)	20(21)	900(16)	.18(4)	.20(6)	1.5(5)	33(17)
27.	Cantaloupe	12(16)	251(23)	17[b](24)	3,400(9)	.04(31)	.03(38)	.6(20)	33(18)
28.	Snap Bean	7(22)	243(26)	32(11)	600(18)	.08(18)	.11(13)	.5(27)	19(26)
29.	Beet	60(4)	335(13)	25(15)	20(40)	.03(38)	.05(26)	.4(29)	10(30)
30.	Celery	126(1)	341(11)	22(19)	240(28)	.03(34)	.03(36)	.3(33)	9(34)
					Crucifers				
31.	Cabbage	20(11)	233(29)	13(29)	130(32)	.05(27)	.05(23)	.3(34)	47(14)
32.	Broccoli	15(12)	382(6)	24(17)	2,500(11)	.10(11)	.23(4)	.9(12)	113(5)
33.	Brussels Sprout	14(13)	390(5)	29(13)	550(21)	.10(13)	.16(9)	.9(13)	102(6)
34.	Cauliflower	13(15)	295(16)	24(16)	60(36)	.11(8)	.10(14)	.7(17)	78(9)
35.	Kale	75(2)	238(7)	37(7)	10,000(2)	.16(6)	.26(3)	2.1(2)	186(1)
36.	Watercress	52(5)	282(17)	20(20)	4,900(8)	.08(17)	.16(8)	.9(11)	79(8)
37.	Mustard green	32(9)	377(8)	27(14)	7,000(7)	.11(9)	.22(5)	.8(15)	97(7)
38.	Turnip green	40[b](7)	250[b](24)	58(3)	7,600(6)	.21(3)	.39(1)	.8(14)	139(3)
39.	Chinese Cabbage	23(10)	253(22)	14(26)	150(31)	.05(25)	.04(28)	.6(19)	25(21)
40.	Rutabaga	5(24)	239(28)	15(25)	580(20)	.07(20)	.07(18)	1.1(8)	43(15)
41.	Collard	40[b](8)	450(3)	57(4)	9,300(3)	.16(5)	.31(2)	1.7(3)	152(2)
42.	Kolhrabi	8(21)	372(9)	37(6)	20(39)	.06(23)	.04(30)	.3(35)	66(10)

Main Source: Agricultural Handbook No. 8, USDA 1963, Washington, D.C.

[a]The Heinz Handbook of Nutrition (1959), McGraw-Hill Book Company, Inc., New York.

[b]Nutrition Composition of Fresh California-Grown Vegetables (1962), California Agricultural Experiment Station Bulletin No. 788, Davis, California. Numbers in parentheses denote relative ranking for each nutritive attribute among 42 fruits and vegetables; 1 = highest and 42 = lowest amount.

TABLE 4

Relative Rankings[a] for Energy, Fat, and Carbohydrates of 42 Fruits and Vegetables Surveyed

Fruits and vegetables	Energy (cal)	Relative ranking	Fat (%)	Relative ranking	Carbohydrates (%)	Relative ranking
Lima bean	123	1	0.5	8	22.1	4
Sweet potato	114	2	0.4	12	26.3	1
Sweet corn	96	3	1.0	2	22.1	3
Banana	85	4	0.2	30	22.2	2
Pea	84	5	0.4	11	14.4	10
Potato	76	6	0.1	42	17.1	6
Grape	69	7	1.0	1	15.7	7
Plum	66	8	0.2[b]	29	17.8	5
Pear	61	9	0.4	10	15.3	8
Apple	58	10	0.5	5	14.5	9
Sour cherry	58	11	0.3	17	14.3	11
Kale*	53	12	0.8	4	9.0	19
Apricot	51	13	0.2	28	12.8	12
Orange	49	14	0.2	27	12.2	13
Rutabaga*	46	15	0.1	41	11.0	14
Other crucifers						
Collard	45	16	0.8	3	7.5	24
Brussels sprout	45	17	0.4	9	8.3	22
Broccoli	32	25	0.3	16	5.9	28
Mustard green	31	26	0.5	7	5.6	29
Kohlrabi	29	28	0.1	36	6.6	26
Turnip green	28	29	0.3	15	5.0	33
Cauliflower	27	30	0.2	25	5.2	31
Cabbage	24	34	0.2	22	5.4	30
Watercress	19	37	0.3	13	3.0	41
Chinese cabbage	14	42	0.1	32	3.0	40

a[1] = highest amount; 42 = lowest amount.
b[Source]: The Heinz Handbook of Nutrition, McGraw-Hill Book Company, Inc., New York.
*Crucifers among top 15 fruits and vegetables.

maintenance and growth depends upon metabolism, age, sex, and other physiological factors. The protein quality is determined in terms of the pattern of essential amino acids in the structure of protein.

To evaluate the status of crucifers with regard to their protein content, one must consider the protein quality in terms of balanced essential amino acids. In addition, one must assess their availability to man. This can be determined by biological evaluation through human and animal feeding experiments. The biological value of proteins expresses the efficiency with which the body utilizes absorbed proteins as a dietary source of amino acids. This biological value may be altered if the body need for one amino acid or group of amino acids changes relative to the requirements of other amino acids.

Protein requirements are based on an involvement of total nitrogen, the pattern of essential amino acids, optimal ratio of essential to nonessential amino acids, and their availability to the body cells.

Other nutrients essential to protein utilization — Any mineral or vitamin present in the diet which is necessary for normal growth and metabolism can be presumed to influence the utilization of dietary protein. A deficiency of essential vitamins or minerals leads to lack of growth and deposition of proteins in tissues. An adequate supply of B-complex vitamins is helpful for the critical protein utilization. A suboptimal intake of the B-complex vitamins may intensify the depressive effect of a low-quality protein diet on growth (or as far as growth is concerned). It is also known that the deficiency of potassium, phosphorus, sodium, calcium, and magnesium influences the body capacity to utilize amino acids and protein.

Quantitative studies on amino acids have established the fact that valine, leucine, isoleucine, methionine, threonine, phenylalanine, tryptophan, and lysine are required in the diet for the maintenance of nitrogen equilibrium and prevention of subjective disease symptoms. Kemmerer and Acosta (1949) showed that the protein of cauliflower, as compared with whole egg protein, was fairly well balanced as far as the essential amino acids were concerned. There was an optimum presence of methionine, valine, threonine, histidine, tryptophan,and arginine. On the other hand, Hirsch et al. (1952) reported that among the ten essential amino acids in broccoli, using the amino acid content of whole egg as a criterion, isoleucine, leucine, and methionine are limiting.

Tables 5, 6, and 7 show that cruciferous vegetables are rich sources of proteins among the fruits and vegetables surveyed. The crucifers contain all the essential amino acids and, particularly, sulfur-containing amino acids. When compared with the richest source of plant proteins, peas and beans, the crucifers compare very favorably in biological value, digestibility, and net protein utilization.

In general, the utilization of vegetable proteins is improved as a result of mild heat treatments. Heat also deactivates various inhibitors and toxic substances which lower the nutritive value of some raw vegetables. In addition, in certain foods a nonenzymatic browning reaction (Maillard) can make a contribution to food technology by producing desired colors and flavors, particularly in dehydrated and canned products. However, excessive heat can damage proteins.

Minerals

The criteria that determine the essentiality of minerals in the diet are determined by the requirement for proper growth and good health. Those minerals needed in macro (major) amounts are calcium, phosphorus, magnesium, potassium,

TABLE 5

Relative Rankings[a] for Protein of 42 Fruits and Vegetables Surveyed

Fruits and vegetables	Protein (%)	Relative rankings	Water (%)
Lima bean	8.4	1	67.5
Pea	6.3	2	78.0
Kale*	6.0	3	82.7
Brussels sprout*	4.9	4	85.2
Collard*	4.8	5	85.5
Broccoli*	3.6	6	89.1
Sweet corn	3.5	7	72.7
Spinach	3.2	8	90.2
Turnip green*	3.0	9	90.3
Mustard green*	3.0	10	89.5
Cauliflower*	2.7	11	91.0
Asparagus	2.5	12	91.7
Watercress*	2.2	13	93.3
Potato	2.1	14	79.8
Kohlrabi*	2.0	15	90.3
Other crucifers			
Cabbage	1.3	20	92.4
Chinese cabbage	1.2	23	95.0
Rutabaga	1.1	29	87.0

[a]1 = highest amount; 42 = lowest amount.
*Crucifers among top 15 fruits and vegetables.

Note: Most cruciferous vegetables contain higher amounts of water than leguminous vegetables. This table indicates that, on the dry weight basis, they contain substantially more protein.

TABLE 6

Amino Acid Content of Cruciferous Vegetables, 100 g per Edible Portion

Vegetables	Trypto-phan (mg)	Threo-nine (mg)	Iso-leucine (mg)	Leucine (mg)	Lysine (mg)	Methio-nine (mg)	Cystine (mg)	Phenyl-alanine (mg)	Tyro-sine (mg)
Brussels sprout (4.4% protein)	44	153	186	194	197	46	(28)[a]	148	–
Cabbage (1.4% protein)	11	39	40	57	66	13	28	30	30
Collard (3.9% protein)	55	114	121	218	202	46	59	124	151
Kale (3.9% protein)	42	139	133	252	121	35	36	158	–
Mustard green (2.3% protein)	37	60	75	62	111	24	36	74	121
Turnip green (2.9% protein)	45	125	107	207	129	52	45	146	105
Watercress (1.7% protein)	28	84	76	131	91	10	–	62	36
Broccoli (3.3% protein)	37	122	126	163	147	50	(48)	119	–
Cauliflower (2.4% protein)	33	102	104	162	134	47	–	75	34
Snap bean[b] (2.4% protein)	33	91	109	139	126	35	24	57	50

Vegetables	Valine (mg)	Arga-nine (mg)	Histi-dine (mg)	Ala-nine (mg)	Aspartic acid (mg)	Glutamic acid (mg)	Glycine (mg)	Proline (mg)	Serine (mg)
Brussels sprout (4.4% protein)	193	279	106	–	–	–	–	–	–
Cabbage (1.4% protein)	43	105	25	71	196	266	(18)	(59)	42
Collard (3.9% protein)	195	258	87	–	–	–	–	–	–
Kale (3.9% protein)	184	202	62	64	–	–	–	–	–
Mustard green (2.3% protein)	108	167	41	–	–	–	–	–	–
Turnip green (2.9% protein)	149	167	51	(164)	(247)	(342)	(136)	(126)	(116)
Watercress (1.7% protein)	84	53	34	60	242	163	–	–	–
Broccoli (3.3% protein)	170	192	63	–	–	–	–	–	–
Cauliflower (2.4% protein)	144	110	48	192	199	185	165	–	–
Snap bean[b] (2.4% protein)	115	101	45	97	(285)	(254)	(90)	(90)	(127)

From Orr, M. L. and Watt, B. K., Amino Acid Content of Foods, Home Economic Research Report No. 4, USDA 1957, Washington, D.C.

[a] Amounts in parentheses were obtained from Amino Acid Content of Foods and Biological Data on Proteins, 1970, Food Policy and Food Science Nutrition Division, FAO, United Nations, Rome, Italy.

[b] Snap bean composition is used for comparison.

Note variations in content of proteins in this table as compared to data presented in Table 5.

TABLE 7

Biological Data[a] of Proteins in Selected Cruciferous Vegetables

Crucifers	Biological Value[b]	Digestibility[c]	NPU (calculated)[d]
Cabbage	39.9	87.8	35.2
Kale	63.9	85.0	54.3
Mustard	73.6	81.8	60.2
Rape	77.9	84.8	66.1
Turnip greens	52.3	86.0	45.1
Pea[e]	65.2	81.0	52.8

[a]From Amino Acid Content of Foods, FAO, 1970.
[b]Biological Value (BV): The proportion of absorbed nitrogen that is retained in the body for maintenance and/or growth.
[c]Digestibility (D): The proportion of food nitrogen that is absorbed.
[d]Net Protein Utilization (NPU) calculated: Obtained from the determined values of the Biological Value and from Digestibility or NPU = BV X D.
[e]Pea is chosen as a reference.

TABLE 8

Relative Rankings[a] for Calcium Content of 42 Fruits and Vegetables Surveyed

Fruits and vegetables	Calcium (mg/100 g)	Relative rankings
Collard*	250	1
Kale*	249	2
Turnip green*	246	3
Mustard green*	183	4
Watercress*	151	5
Broccoli*	103	6
Spinach**	93	7
Rutabaga*	66	8
Snap bean	56	9
Lima bean	57	10
Cabbage*	49	11
Chinese cabbage*	43	12
Kohlrabi*	41	13
Orange	41	14
Celery	39	15
Other crucifers		
Brussels sprout	36	17
Cauliflower	25	24

[a]1 = highest amount; 42 = lowest amount.
*Crucifers among top 15 fruits and vegetables.
**Calcium in spinach is largely unavailable.

sulfur, sodium, and chlorine, and those needed in micro (trace) amounts are iron, iodine, copper, cobalt, chromium, manganese, selenium, zinc, fluorine, and molybdenum.

Kelly (1943) compared the calcium content of cabbage and turnip greens with milk; he found both vegetables are good sources of calcium with relative availability of 93% and 87%, respectively.

Oberg, et al. (1946) reported that turnip greens contain significant amounts of calcium and iron.

All of the cruciferous vegetables are excellent sources of minerals such as calcium, iron, magnesium, sodium, potassium, and phosphorus as compared with 30 other fruits and vegetables surveyed. Most of these minerals are in available form (Tables 8, 9, 10, 11, 12, and 13).

TABLE 9

Relative Rankings[a] for Phosphorus Content of 42 Fruits and Vegetables Surveyed

Fruits and vegetables	Phosphorus (mg/100 g)	Relative rankings
Lima bean	142	1
Pea	116	2
Sweet corn	111	3
Kale*	93	4
Collard*	82	5
Brussels sprout*	80	6
Broccoli*	78	7
Asparagus	62	8
Turnip green*	58	9
Cauliflower*	56	10
Watercress*	54	11
Potato	53	12
Kohlrabi*	51	13
Spinach	51	14
Mustard green*	50	15
Other crucifers		
Chinese cabbage	40	18
Rutabaga	39	19
Cabbage	29	24

[a]1 = highest amount; 42 = lowest amount.
*Crucifers among top 15 fruits and vegetables.

TABLE 10

Relative Rankings[a] for Iron Content of 42 Fruits and Vegetables Surveyed

Fruits and vegetables	Iron (mg/100 g)	Relative rankings
Spinach**	3.1	1
Mustard green*	3.0	2
Lima bean	2.8	3
Kale*	2.7	4
Lettuce	2.0	5
Pea	1.9	6
Turnip green*	1.8	7
Watercress*	1.7	8
Collard*	1.5	9
Brussels sprout*	1.5	10
Cucumber	1.1	11
Cauliflower*	1.1	12
Broccoli*	1.1	13
Asparagus	1.0	14
Strawberry	1.0	15
Other crucifers		
Chinese cabbage	0.6	23
Kohlrabi	0.5	27
Cabbage	0.4	33
Rutabaga	0.4	36

[a]1 = highest amount; 42 = lowest amount.
*Crucifers among top 15 fruits and vegetables.
**Iron content in spinach is largely unavailable.

TABLE 11

Relative Rankings[a] for Sodium Content of 42 Fruits and Vegetables Surveyed

Fruits and vegetables	Sodium (mg/100 g)	Relative rankings
Celery	126	1
Kale*	75	2
Spinach	71	3
Beets	60	4
Watercress*	52	5
Carrot	47	6
Turnip green[b]*	40	7
Collard[b]*	40	8
Mustard green*	32	9
Chinese cabbage*	23	10
Cabbage*	20	11
Broccoli*	15	12
Brussels sprout*	14	13
Pepper	13	14
Cauliflower*	13	15
Other crucifers		
Kohlrabi	8	21
Rutabaga	5	24

[a]1 = highest amount; 42 = lowest amount.
[b]From Nutrient Composition of Fresh California Grown Vegetables, California Agricultural Experiment Station, Bulletin No. 788, Davis, California.
*Crucifers among top 15 fruits and vegetables.

TABLE 12

Relative Rankings[a] for Potassium Content of 42 Fruits and Vegetables Surveyed

Fruits and vegetables	Potassium (mg/100 g)	Relative rankings
Lima bean	650	1
Spinach	470	2
Collard*	450	3
Potato	407	4
Brussels sprout*	390	5
Broccoli*	382	6
Kale*	378	7
Mustard green*	377	8
Kohlrabi*	372	9
Banana	370	10
Celery	341	11
Carrot	341	12
Beet	335	13
Pea	316	14
Plum	299	15
Other crucifers		
Cauliflower	295	16
Watercress	282	17
Chinese cabbage	253	22
Turnip green[b]	250	24
Rutabaga	239	28
Cabbage	233	29

[a]1 = highest amount; 42 = lowest amount.
[b]From Nutrient Composition of Fresh California Grown Vegetables, California Agricultural Experiment Station, Bulletin No. 788, Davis, California.
*Crucifers among top 15 fruits and vegetables.

TABLE 13

Relative Rankings[a] for Magnesium Content of 42 Fruits and Vegetables Surveyed

Fruits and vegetables	Magnesium (mg/100 g)	Relative rankings
Spinach	88	1
Lima bean	67	2
Turnip green*	58	3
Collard*	57	4
Sweet corn	48	5
Kohlrabi*	37	6
Kale*	37	7
Pea	35	8
Potato	34	9
Banana	33	10
Snap bean	32	11
Sweet potato	31	12
Brussels sprout*	29	13
Mustard green*	27	14
Beet	25	15
Other crucifers		
Cauliflower	24	16
Broccoli	24	17
Watercress	20	20
Rutabaga	15	25
Chinese cabbage	14	26
Cabbage	13	29

[a]1 = highest amount; 42 = lowest amount.
*Crucifers among top 15 fruits and vegetables.

Vitamins

Vitamins are a group of organic compounds that are essential in relatively minute quantities for the metabolism of other nutrients in the body for normal growth, maintenance of health, and reproduction. They prevent diseases and participate in the regulation of the body processes. The Food and Nutrition Board of the National Research Council (1968) has recommended that a specific amount of the 13 known vitamins be included in the diet of the adult each day. These include the fat soluble vitamins—A, β-carotene, D, E (a-Tocopherol), and K; and water soluble—ascorbic acid, biotin, folacin, niacin, pantothenic acid, riboflavin, thiamine, B_6, and B_{12}. Cruciferous vegetables contain most of these in significant amounts.

Clayton and Borden (1943), by means of blood and urine analyses, found that the ascorbic acid content of raw cabbage was completely utilized.

Tables 14, 15, 16, 17, and 18 show that cruciferous vegetables have substantial amounts of vitamins A (carotene), C, riboflavin, niacin, and thiamine. Information on folic acid, pantothenic acid, and pyridoxine is limited (Table 19).

In general, it seems from the summary table (Table 20) obtained from data presented in Table 3 by rank correlation analysis, that the cruciferous vegetables are the most efficient in synthesizing high concentrations of proteins, amino acids, minerals, and vitamins and are low in caloric content.

The comparison of relative nutritive value and production data shows that Americans are cultivating and consuming very few of the vegetables that are best for them. Generally, cabbage, cauliflower, broccoli, Brussels sprouts, turnip greens, mustard greens, and kale, which rank high in nutritive value, are disliked and the least consumed.

Flavor and Preference Considerations
Cruciferous Flavors

Along with nutritional attributes, flavor is one of the most important properties of food. It is the flavor which determines whether we like or dislike

TABLE 14
Relative Rankings[a] for Vitamin A Content of 42 Fruits and Vegetables Surveyed

Fruits and vegetables	Vitamin A (IU)	Relative rankings
Carrot	11,000	1
Kale*	10,000	2
Collard*	9,300	3
Sweet potato	8,800	4
Spinach	8,100	5
Turnip green*	7,600	6
Mustard green*	7,000	7
Watercress*	4,900	8
Cantaloupe	3,400	9
Apricot	2,700	10
Broccoli*	2,500	11
Peach	1,300	12
Sour cherry	1,000	13
Lettuce	900	14
Tomato	900	15
Other crucifers		
Rutabaga	580	20
Brussels sprout	550	21
Chinese cabbage	150	31
Cabbage	130	32
Cauliflower	60	36
Kohlrabi	20	39
Orange	200	29

[a]1 = highest amount; 42 = lowest amount.
*Crucifers among top 15 fruits and vegetables.

TABLE 15
Relative Rankings[a] for Thiamine Content of 42 Fruits and Vegetables Surveyed

Fruits and vegetables	Thiamine (mg/100 g)	Relative rankings
Pea	0.35	1
Lima bean	0.24	2
Turnip green*	0.21	3
Asparagus	0.18	4
Collard*	0.16	5
Kale*	0.16	6
Sweet corn	0.15	7
Cauliflower*	0.11	8
Mustard green*	0.11	9
Spinach	0.10	10
Broccoli*	0.10	11
Orange	0.10	12
Brussels sprout*	0.10	13
Potato	0.10	14
Sweet potato	0.10	15
Other crucifers		
Watercress	0.08	17
Rutabaga	0.07	20
Kohlrabi	0.06	23
Chinese cabbage	0.05	25
Cabbage	0.05	27

[a]1 = highest amount; 42 = lowest amount.
*Crucifers among top 15 fruits and vegetables.

TABLE 16

Relative Rankings[a] for Riboflavin Content of 42 Fruits and Vegetables Surveyed

Fruits and vegetables	Riboflavin (mg/100 g)	Relative rankings
Turnip green*	0.39	1
Collard*	0.31	2
Kale*	0.26	3
Broccoli*	0.23	4
Mustard green*	0.22	5
Asparagus	0.20	6
Spinach	0.20	7
Watercress*	0.16	8
Brussels sprout*	0.16	9
Pea	0.14	10
Sweet corn	0.12	11
Lima bean	0.12	12
Snap bean	0.11	13
Cauliflower*	0.10	14
Squash	0.09	15
Other crucifers		
Rutabaga	0.07	18
Cabbage	0.05	23
Chinese cabbage	0.04	28
Kohlrabi	0.04	30

[a]1 = highest amount; 42 = lowest amount.
*Crucifers among top 15 fruits and vegetables.

TABLE 17

Relative Rankings[a] for Niacin Content of 42 Fruits and Vegetables Surveyed

Fruits and vegetables	Niacin (mg/100 g)	Relative rankings
Pea	2.9	1
Kale*	2.1	2
Collard*	1.7	3
Sweet corn	1.7	4
Asparagus	1.5	5
Potato	1.5	6
Lima bean	1.4	7
Rutabaga*	1.1	8
Squash	1.0	9
Peach	1.0	10
Watercress*	0.9	11
Broccoli*	0.9	12
Brussels sprout*	0.9	13
Turnip green*	0.8	14
Mustard green*	0.8	15
Other crucifers		
Cauliflower	0.7	17
Chinese cabbage	0.6	19
Kohlrabi	0.3	35
Cabbage	0.3	34

[a]1 = highest amount; 42 = lowest amount.
*Crucifers among top 15 fruits and vegetables.

TABLE 18

Relative Rankings[a] for Vitamin C of 42 Fruits and Vegetables Surveyed

Fruits and vegetables	Vitamin C (mg/100 g)	Relative rankings
Kale*	186	1
Collard*	152	2
Turnip green*	139	3
Pepper	128	4
Broccoli*	113	5
Brussels sprout*	109	6
Mustard green*	97	7
Watercress*	79	8
Cauliflower*	78	9
Kohlrabi*	66	10
Strawberry	59	11
Spinach	51	12
Orange	50	13
Cabbage*	47	14
Rutabaga*	43	15
Other crucifers		
Chinese cabbage	25	21

[a]1 = highest amount; 42 = lowest amount.
*Crucifers among top 15 fruits and vegetables.

TABLE 19

Contents of Folic Acid, Pantothenic Acid, and Pyridoxine in Some Fresh Cruciferous Vegetables

Crucifers	Folic acid[a] (mg/100 g)	Pantothenic acid[b] (mg/100 g)	Pyridoxine[b] (mg/100 g)
Broccoli[c]	.0352	1.170	.195
Brussels sprout	.0189	.723	.230
Cabbage[c]	.0120[d]	.205	.160
Chinese cabbage	.0114	—	—
Cauliflower[c]	.0172	1.000	.210
Collard	.1018	.450[e]	.195[e]
Kale	.0595	1.000	.300
Kohlrabi	.0101	.165	.150
Mustard green	.0377	.210	—
Rutabaga	.0069	.160	.100
Turnip green[c]	.0695	.380	.263
Watercress	.0467	.310	.129

[a]Determined by biological assay with *Lactobacillus casei*; data taken from Agriculture Handbook No. 29, USDA, 1951.
[b]Data taken from Home Economics Research Report No. 36, USDA, 1969.
[c]Vegetables were trimmed for the analysis of pantothenic acid and pyridoxine.
[d]Determined by biological assay with *Streptococcus faecalis.*
[e]Frozen samples.

TABLE 20

Relative Nutritional Value of 42 Fruits and Vegetables Surveyed for Cumulative of 15 Nutrients

Fruits and vegetables	Rank for relative nutritional value	Water %
Kale*	1	82.7
Collard*	2	85.3
Lima bean	3	67.5
Turnip green*	4	90.3
Pea	5	78.2
Mustard green*	6	89.3
Brussels sprout*	7	85.2
Spinach	8	90.7
Broccoli*	9	89.1
Sweet corn	10	72.7
Watercress*	11	93.3
Sweet potato	12	70.6
Asparagus	13	91.7
Cauliflower*	14	91.0
Snap bean	15	91.1
Other crucifers		
Rutabaga	18	87.0
Kohlrabi	19	90.3
Cabbage	23	92.4
Chinese cabbage	24	95.0
Other major fruits and vegetables		
Potato	17	79.8
Tomato	27	93.5
Orange	29	86.0
Apple	41	84.4
Watermelon	42	92.6

*Crucifers among top 15 fruits and vegetables.
If expressed on dry weight basis, the ranking could be improved in favor of crucifers because of their higher water content.

a particular food. The flavor causes either satisfaction or a negative reaction, with the latter being the subject of many gastronomical anecdotes.

The flavor of cruciferous vegetables depends upon the cultivar, maturity, season, horticultural practices, storage, handling, and method of cooking. According to Johnson et al. (1971a), Konig and Kracht (1921) reported the presence of hydrogen sulfide and methanethiol in cabbage. Jensen et al. (1953), using paper chromatography, showed the presence of 2-propenyl isothiocyanate in both white and red cabbage and 3-butenyl isothiocyanate in red cabbage. Clapp et al. (1959) detected four thioureas from volatile isothiocyanates. Bailey et al. (1961) used gas chromatography and mass spectrometry to examine fresh, dehydrated, and rehydrated cabbage by vacuum steam distillation and fractionation and identified 24 sulfur compounds. Clapp et al. (1959) and Hewitt et al. (1956) reported that the volatile isothiocyanates, important in the flavor of fresh cabbage, could be regenerated by enzymatic hydrolysis from the corresponding thioglucosides in the dehydrated cabbage.

Dateo et al. (1957) noted that cabbage contains thermolabile sulfur compounds which evolve sulfurous odors. The major volatile sulfur compounds (dimethyl disulfide and hydrogen sulfide) of cooked cabbage are derived from a precursor L-S-methylcysteine sulfoxide, a free amino acid. Maruyama (1970) indicated that dimethyl trisulfide was a major aroma component in cooked *Brassicaceous* vegetables. He proposed that the strong, unpleasant aroma that is characteristic of over-cooked *Brassicaceous* vegetables is due to the gradual loss of pleasant volatile components and resultant unmasking of unpleasant sulfur components. This concurs with the findings of Schwimmer (1963) who showed that in cabbage, horseradish, and broccoli the enzyme systems involved in desirable flavor production are less stable than those which give rise to deleterious flavor. Bailey et al. (1961) showed that fresh cabbage contained allyl isothiocyanate, which could be regenerated by enzymatic action in dehydrated cabbage. The identification of five isothiocyanates revealed the absence of corresponding thioglucosides from which the isothiocyanates are released by enzymatic hydrolysis.

MacLeod and MacLeod (1970a and 1970b), assessing the close relation among flavors of cabbage, cauliflower, and Brussels sprouts, found that isothiocyanates with allyl isothiocyanate comprised over 6% of the total volatiles. Brussels sprouts have a stronger flavor (five times more allyl cyanide) than cabbage. Chopping increases the release of dimethyl sulfide in both vegetables. However, the flavor profile of Brussels sprouts differs slightly from that of cabbage, as chopping sprouts reduces their production of carbonyls and increases that of thioglucoside sinigrin (allyl cyanide and allyl isothiocyanate). They further reported that dehydrated cabbage possessed a strong and objectionable odor and contained eight or more times as much volatile material as fresh cabbage, the increase coming from bland or objectionable compounds (saturated aldehydes and 2-propenyl cyanide). Some desirable compounds (2-propenyl isothiocyanate) had virtually disappeared. Storage of dehydrated, cooked cabbage brought other changes, the loss of about half the 2-propenyl cyanide being an advantage.

Nabors and Salunkhe (1969) found that low-boiling compounds of sauerkraut were destroyed faster when dehydration was with conventional or microwave drying than with freeze-drying. Self et al. (1963) identified sulfur-containing compounds — hydrogen sulfide, methanethiol, ethanethiol, propanethiol, and dimethyl sulfide — in addition to acetaldehyde and 2-methyl propanol in cooked cauliflower. Hing and Weckel (1964) in their studies noted that rutabaga flavor consisted of ammonia, isothiocyanates, hydrogen sulfide, dimethyl sulfide, dimethyl disulfide, 2-phenethyl isothiocyanate, mercaptans, and acetaldehyde. According to Johnson et al. (1971b), Hofmann in 1874 found phenyl propionitrile as a major component of watercress and Gadamer in 1899 reported that by grinding the leaves and liberating the enzymes, a different oil was produced that had a strong mustard oil odor and contained phenethyl isothiocyanate. Thus, to date, the compounds obtained from watercress are phenyl propionitrile and phenethyl isothiocyanate. Fariis and Kjaer (1966) showed that the pungent principle in radishes was 4-methylthio-*trans*-3-butenyl isothiocyanate.

Food Preference

An excellent survey was conducted by Einstein and Hornstein (1970) on the food preferences using 207 food items and nearly 50,000 college students representing about 1% of the college enrollment of the United States in 1966-67. The responses were collected on a regional basis, subdivided as to sex, and reported as national and regional totals. The foods were ranked in terms of "liked," "disliked," and "do not know." Median and percentile for each food class as well as for all foods were calculated. The relationships between food preferences and nutritional values were examined. The percent of the recommended dietary allowance for vitamins A and C and minerals, calcium and iron, provided by one serving was calculated for each food. The results indicated that if food preferences were the sole determinant of food intake, then the dietary intake would be low in vitamin A. Cooked cabbage, cauliflower, Brussels sprouts, broccoli, turnips, and kale were the least liked foods. Coleslaw, with its raw cabbage, was an exception. Similarly, Kamen et al. (1967) and Peryam and Seaton (1962), in surveys for the U.S. Army Quartermaster Corps of the food preferences of service men, found that vegetables in general and crucifers in particular were disliked. Van Riter (1956) made a study in a college women's dormitory and found that strong-flavored vegetables such as turnips, broccoli, and cabbage were the least acceptable among 26 vegetables surveyed.

According to the Household Food Consumption Survey in 1965-66, (USDA 1972) on the dark-green and deep-yellow vegetable group, only 10 to 20% of all the individuals contacted ate them. The consumption was significantly higher in the South than in the North, in the urban than in the rural areas, and in the under $3,000 than in the over $8,000 annual income class. The percentage of consumers eating these vegetables in restaurants was very low (about 4%). More people (15 to 40%) had diets below recommended allowances for calcium and iron than for other nutrients. In addition, these individuals ate significantly less (14 to 45%) vitamin A, thiamine, riboflavin, ascorbic acid, vitamin B_6, magnesium, and vitamin B_{12} than was required.

In general, cabbage, kale, cauliflower, broccoli, Brussels sprouts, mustard greens, turnips, rutabaga, etc., which contain high amounts of objectionable sulfur compounds, have a mushy texture after extended boiling. They also undergo significant loss of color from prolonged heating in the kitchens of restaurants or dormitories, and are not preferred by the consumers despite their being excellent sources of protein, vitamins and minerals, and having few calories. Hence it becomes a challenge to make these vegetables more acceptable by reducing the sulfur compounds without losing the sulfur-containing amino acids while still maintaining crisp texture and attractive color during processing and prior to serving. Research must be geared to this end if these highly nutritious and low-caloried vegetables are to be made into more acceptable products. Market research has shown that the key factor in the success of new products or new forms of old products is not cost, nutrition, convenience, nor stability, it is the QUALITY as shown by the flavor, texture, and color of the food served at the table.

Toxicological Considerations

Vegetables of the cruciferae family in general, and of the *Brassica* genus in particular, contain goitrogens — antinutrients — which cause enlargement of the thyroid glands. Natural thioglucosides (glucosinolate) are the source of goitrogens. The same thioglucosides with their associated enzyme(s) impart the desirable culinary flavor of cabbage, broccoli, and cauliflower. Kjaer (1960) indicated that the synthesis of allylthiocyanate is due to the activity of the enzyme systems. Thioglucosides (nongoitrogens), upon enzymatic hydrolysis, yield glucose and bisulfate.

For example, a thioglucoside — allyl thioglucoside (sinigrin) — is present in cabbage, kale, Brussels sprouts, broccoli, cauliflower, and mustard. When these vegetables are chopped a specific enzyme, thioglucosidase (myrosinase), hydrolyses allyl thioglucoside into glucose, potassium bisulfate, and allylthiocyanate — a goitrogenic compound (Figure 3). In another case, progoitrin or epi-progoitrin, which is responsible for the typical flavor of kale, rape, turnip, rutabaga, and kohlrabi, upon hydrolysis, subsequent to chopping by thioglucosidase, yields glucose, potassium bisulfate, and highly unstable intermediate forming three compounds; namely, thiocyanate, nitrile + sulfur, and goitrin which are goitrogenic (Figure 4). Goitrin (5-vinyloxazolidine-2-thione) is a potent thyrotoxin and is formed through cyclization of an unstable isothiocyanate-containing hydroxyl group.

The thyroid-inhibiting effect of goitrins is due to inhibition of organic binding of iodine. Such action is consistent with observations that this type of goiter is not alleviated by increasing iodine in thyroids. However, thiocyanate, isothiocyanate, and nitrile ions act as goitrogens only when the

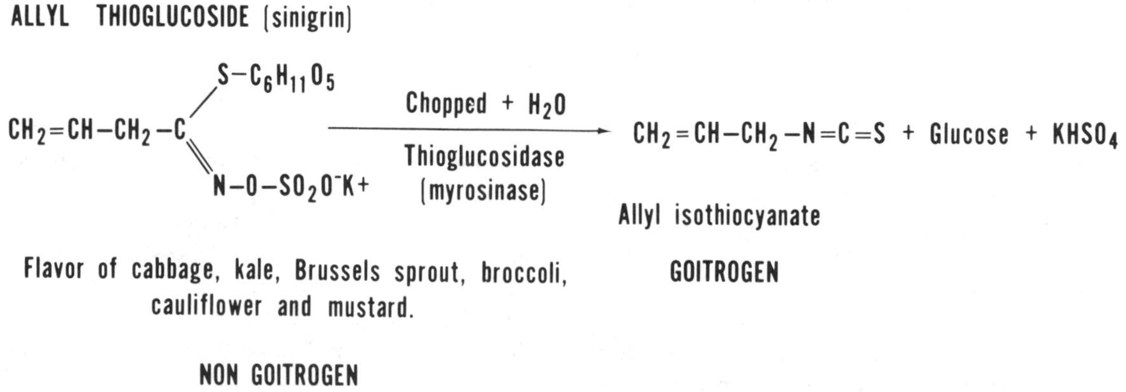

ALLYL THIOGLUCOSIDE (sinigrin)

Flavor of cabbage, kale, Brussels sprout, broccoli, cauliflower and mustard.

NON GOITROGEN

Allyl isothiocyanate

GOITROGEN

FIGURE 3. The enzymatic hydrolysis of allyl thioglucoside (Sinigrin) to allyl isothiocyanate.

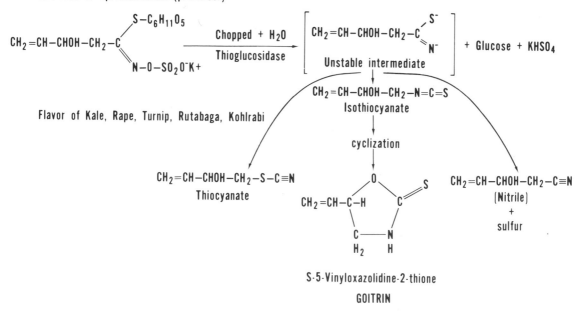

PROGOITRIN or epi-PROGOITRIN (precursor)

FIGURE 4. The enzymatic hydrolysis of progoitrin – a thioglucoside – to goitrin.

iodine content in the diet is low. In regions where the iodine content of the diet is low, benign goiter may be accentuated by eating excessive amounts of *Brassica* vegetables.

Crosby (1966) and Orgell et al. (1959) reported that aqueous and ethyl acetate extracts from leaves of broccoli, cabbage, rutabaga, turnips, and radishes inhibit human plasma cholinesterase. This indicates that such extracts contain chemicals that may modify the functions of the nervous system. Identification and determination of the amount of each goitrogen and cholinesterase inhibitor in cruciferous plants including the biological assaying of each substance need to be studied. The flatulence distress syndrome after eating cooked crucifers is not as chronic and not as offensive as that produced after the consumption of beans, sweet potatoes, and onions (Murphy, 1973).

Greer (1956) found that most of the goitrogen properties of the product might be lost during cooking. Greer (1962) also estimated that iodine deficiency is the cause of endemic nontoxic goiter in all but 4% of the cases. Of this 4%, it cannot be demonstrated that goitrogens in *Brassicaceous* vegetables are the major causal agents. Two cruciferous vegetables having high concentrations of thiocyanate are cauliflower and kale. However, a daily intake of about 22 lb of cauliflower or kale would be required to furnish a goitrogenic concen-

tration of thiocyanate in the blood. Nevertheless, these thyrotoxic substances should be a concern of the food technologist in developing new or improved processing methods for reasons of economy, convenience, nutrition, flavor, or esthetic appeal. In processing foods from crucifers by new methods, it is essential to insure that goitrogenic substances are not concentrated in the product.

Therapeutic and Other Pharmacological Properties

Cheney (1942, 1949, 1950, 1952, 1954, 1955) and Cheney et al. (1956) of the Stanford Medical School reported that a new entity which they termed vitamin U is an important constituent of cabbage. In a series of articles, he and his co-workers reviewed the raw cabbage juice therapy for peptic ulcers and indicated that an antipeptic ulcer dietary factor is present in fresh cabbage juice. This factor is readily destroyed by heat. A clinical study of patients indicated that concentrated cabbage juice is significantly effective in healing peptic ulcers. An experimental preparation MK-72 has been prepared by Merck and Company. Steigmann and Shulman (1952) and Strehler and Hunziker (1954) have devoted further study to this antiulcer factor. McRorie et al. (1954) isolated a heat labile analogue of methionine from cabbage juice that parallels the occurrence of vitamin U.

Petering et al. (1948) indicated that cabbage contains "antistiffness" factors.

Lichtenstein et al. (1962, 1964) reported that cruciferous plants containing 2-phenyl ethyl isothiocyanate have insecticidal properties with no apparent harmful effects on humans. From cabbage, Holley et al. (1951) obtained indoleacetic acid, a plant growth hormone, and two other fractions that may be the precursors of indoleacetic acid.

Calloway and Munson (1961) found a substance in cabbage and broccoli that reduces the radiosensitivity of guinea pigs exposed to ionizing radiations. From histological and hematological studies, they concluded that cabbage and broccoli in the guinea pig diet resulted in improved nutrition and larger livers with a higher storage of vitamin A and mineral and fat content. Calloway et al. (1963) further stated that these diets supplemented with broccoli, which has a high content of β-carotene, when fed for two weeks before exposure to X-radiation or gamma radiation, consistently and significantly reduced mortality.

Levine et al. (1932) showed that nutritional anemia was controlled by consumption of broccoli and turnip greens and concluded that these vegetables are excellent sources of available iron and copper for the formation of hemoglobin. Pederson and Fisher (1944), Foter (1940), and Foter and Golick (1938) showed that allyl isothiocyanate and phenyl ethyl isothiocyanate components of crucifers have bactericidal properties. According to Pederson and Albury (1969), Letzig stated that sauerkraut and its juice are laxative. Nabb and O'Dell (1964) reported that fresh and dehydrated cabbage increased *Salmonella* resistance in the guinea pig. This unidentified protective factor was water insoluble. Lakhanpal et al. (1966) and Singh et al. (1968) showed that cabbage and broccoli stimulated the growth rate of guinea pigs and protected them against *Salmonellosis*. The growth stimulating effect was destroyed by drying, but the *Salmonellosis*-resistance factor was still present.

Processing and Storage Considerations

Processing is difficult to define. Perhaps by defining it, we limit the scope of the "processing" process. However, for the sake of this presentation, we would like to include those treatments which begin after harvest and end

before consumption: handling, transportation, refrigeration, holding, washing, trimming, blanching, freezing, canning, drying, packaging, chemicals, radiation, storage, and ultimately cooking or heating.

Changes in the composition of raw vegetables which may decrease their nutritive value can occur after harvest, during transportation, holding, handling, processing, and subsequent storage and distribution. After they are harvested, fruits and vegetables are still physiologically active. Enzymatic and respiratory processes may bring about profound changes unless controlled. In general, the lower the temperature at which a commodity is stored, the less likely it is to develop abnormalities with resultant loss of quality and of nutritive value. The extent of these losses depends upon the type of vegetable, its anatomical and morphological structure, rate of respiration, and the content of respiratory substrates such as sugars and organic acids, and external conditions.

The food technologist and nutritionist must better understand the behavior of produce during its growth. They should know how it changes in composition as it develops, matures, and is stored before processing, and subsequent storage. Then they can predict nutritional alterations during handling, processing, and subsequent storage.

Raw Material Handling: Washing, Trimming, and Blanching

Leafy vegetables may be tumbled in a drum-type cleaner to dislodge sand and soil particles, but these mechanical cleaning operations can bruise or crush delicate tissues and increase respiration and oxidation of ascorbic acids. Unfortunately, literature is extremely limited regarding changes in the composition of vegetables brought about during washing and trimming for commercial processing. The losses that occur during such processes depend upon the crop, cultivar, maturity, freshness, and season. When leafy vegetables are trimmed, the nutrient losses generally exceed the weight losses because nutrients tend to concentrate in the outer leaves. Sheets et al. (1941) reported that outer cabbage leaves contain two to four times more iron than the stem. There is 1.5 to 3 times as much iron, 1.5 times as much ascorbic acid, and 21 times as much carotene in the outer green cabbage leaves as in the inner bleached leaves.

Loss of moisture from leafy vegetables by

evapotranspiration occurs during holding and transporting. Washing in cold water tends to quickly rehydrate such produce.

Water Blanching

The meager published data suggest that washing or other contact with water after blanching can remove an appreciable amount of water-soluble nutrients from leafy vegetables. The objectives of blanching vary with the product and method of preparation to be employed. In freezing vegetables, the inactivation of enzymes is essential because no further heat is employed. In canning, the aim is to expel gases that might contribute to oxidation and to internal pressure from inter-cellular spaces of the vegetables. These examples indicate the need to evaluate blanching methods on the basis of their functions in commercial preserva-tion as well as for the changes they produce in the nutritive composition of the vegetable and the effects on retention of the more thermal-stable niacin, riboflavin, and thermal-labile thiamine during blanching. The influence of blanching on the degradation of the nutritive composition varies from product to product. High temperature-short time blanching in steam is more conducive to the retention of water-soluble nutrients. Certain commercial blanching practices could be modified to higher temperatures and shorter times to improve nutrient retention without loss in quality.

Published information is meager on the effects of blanching operations employed by commercial processors and on the nutritional composition of vegetables, particularly crucifers. It is reasonable to assume that the type of equipment, blanching method, time, temperature, maturity, and cultivar of the produce would influence the degree to which the chemical composition changes.

Sulfite or sulfur dioxide is often added to blanching water to improve the stability of the vegetables used for dehydration or freezing. Mallette et al. (1946) reported that sulfite in blanching water improved the ascorbic acid reten-tion in cabbage, but most of the thiamine was destroyed.

Microwave Blanching

Broccoli was blanched with electronic energy by heating with 3000-mc radiation (radar) (Proctor and Goldblith, 1948). The broccoli in plastic bags was heated for 20 to 30 sec which was sufficient to inactivate catalase and peroxidase. It

was then cooled by placing the bags in cold water. They further reported that under these conditions there was a 100% ascorbic acid retention. In contrast, only 37 to 100% was retained in steam and 24 to 93% in boiling water, where the losses resulted from leaching rather than destruction. Blanching dielectrically in a closed container tended to conserve more ascorbic acid and soluble solids than blanching in water or in steam. The methods for subsequent cooling with a water dip, water spray, or air current were as important as the method of heating. As the vegetables were in closed containers and not in contact with water, certain other functions of the blanching such as the reduction of microorganisms and the removing of mucilaginous materials that influence flavor or appearance of the finished product were not achieved. Dietrich et al. (1970) compared micro-wave, conventional, and combination blanching of Brussels sprouts and found that combination blanching by microwave energy and steam or water effectively inactivated peroxidase enzyme and maintained stability of flavor. It also retained chlorophyll and ascorbic acid as well or better than with conventional blanching and stored at -20, 0, and $20°F$. However, Eheart (1967) report-ed that conventional blanching was superior to microwave blanching, retaining chlorophyll and ascorbic acid in broccoli. Additional information is needed before dielectric blanching becomes a com-mercial reality. In some products, the ascorbic acid seems relatively stable. Perhaps the presence of glucosides, phenolic compounds, or other sub-stances may be responsible for this protection.

Water Quality

The quality of the water used for processing influences the quality and nutritive value of the product. When hard water containing calcium and magnesium is used for canning and freezing, polygalacturonic acid or demethoxylated pectin combines with calcium and magnesium to yield calcium and magnesium pectates. These com-pounds lend additional firmness to the tissue and consequently give a better texture. However, some salts and metals are detrimental to the nutritive value and flavor of the processed products. Natural and industrial pollutants can make culinary water unfit for food processing. Sulfur compounds in water cause detinning and, subsequently, corrosion of the container; tin may bleach the product; and polymerization of certain compounds such as carbonates, chlorides, and sulfates during processing and after storage can precipitate and form scum and cloudiness. Iron and copper compounds combine with tannins and cause blackening and may accelerate the degradation of ascorbic acid. Sodium, magnesium, and calcium sulfates impart a bitter flavor to the processed products. Certain salts of zinc, cadmium, and chromium have toxic effects (Salunkhe et al., 1971a, and Chiang et al., 1971).

All these constituents affect the quality, nutritive value, and wholesomeness of the processed products; consequently, they could cause considerable economic losses to industry and cause the consumers to receive products of inferior quality and low-nutritive value. To meet Public Health Standards and to manufacture products of high quality and nutritive value, processing water must not contain excessive amounts of the impurities (Wilson, 1971).

Salad Preparations: Chopping, Mincing, Grating, and Dressing

Munsell et al. (1949) reported losses of 52 or 19% of the reduced ascorbic acid content when cabbage was minced or shredded but losses in total ascorbic acid were only 3 and 6%, respectively. No further losses in total or reduced ascorbic acid occurred when shredded cabbage was held 3 or 22 hr or minced cabbage for 3 hr. However, minced cabbage showed further losses when held 22 hr. Coleslaw treated with French dressing and held for 22 hr showed still greater loss. Wilcox and Neilson (1947) found that the greatest loss (10%) of ascorbic acid occurred during the shredding or chopping of cabbage. Whether it was added with or without salad dressing, there was a slight loss. However, the addition of salad dressing increased the dehydroascorbic acid twofold in shredded cabbage and threefold in chopped cabbage. Clayton and Goos (1947) observed a 53% loss in 2 hr when vinegar was used. Quinn et al. (1946) showed that, at room temperature, French dressing, vinegar, mayonnaise, and lemon-juice mayonnaise aided in the retention of ascorbic acid in raw, shredded cabbage salad. The lemon-juice mayonnaise retained the most, vinegar mayonnaise the next, and French dressing the least. Wood et al. (1946) reported losses of 9 to 15% in ascorbic acid, 65% in thiamine, and 0 to 3% in riboflavin in diced cabbage with no appreciable further loss when held at room temperature for 2 hr. Francis

(1960) reported that the discoloration of coleslaw was due to oxygen in the container. Tightly closed 300 MSAD cellophane of 1-mil polyethylene bags, when stored below 50°F, effectively preserved the color. The treatment of coleslaw with oxytetracycline delayed the onset of fermentation and deterioration. According to Harris and Von Loesecke (1971), Ashikaga and Chachi in 1951 reported that grated radishes lost 27% thiamine in 24 hr at 50°F.

Refrigerated Products

Haller (1947) noted that roots of turnips when trimmed, washed, and waxed stored longer and had better quality. Washing turnips improved appearance and reduced decay. Waxing reduced weight losses and produced a glossy appearance. Zepplin and Elvehjem (1944) reported that the ascorbic acid in broccoli was destroyed rapidly at room temperature, but it was significantly retained under refrigeration. Other factors such as relative humidity, temperature, exposure to air, and the physical condition of the broccoli were also important in the preservation of ascorbic acid. Ezell and Wilcox (1959) found that the wilting of cabbage, cauliflower, kale, collard, and turnip greens appreciably reduced the ascorbic acid contents. Hence, control of temperature and relative humidity under refrigeration conditions is essential. They also reported in 1962 that temperature and humidity are primary factors in preserving carotene in kale, collard, and turnip greens. Wilting resulted in rapid loss of carotene. Eheart and Odland (1972) showed that when fresh broccoli and green beans were stored at 36°F for 7 days, broccoli lost no ascorbic acid, but green beans lost 88%. This difference may be due to the presence of the sulfhydryl group in the broccoli. Fager et al. (1949) found that the destruction of folic acid of fresh leafy vegetables can be reduced by storing them under refrigeration. Olsen et al. (1947) reported that the folic acid content of broccoli, cauliflower, kohlrabi, radishes, and cabbage was retained better when stored under the refrigeration temperatures. The losses were large when these vegetables were stored at room temperature. Stevens (1943) observed that the nutritive value of dehydrated cabbage and rutabaga was influenced by cultivar, conditions of production, condition of vegetables, blanching, drying, storage, method of reconstitution, and activity of enzymes present.

Lyons and Rappaport (1959) reported that the respiration rate of Brussels sprouts was lower (nearly one fifth) at refrigeration (32°F) than at room temperature (68°F) and satisfactory organoleptic quality was maintained for 72 days at 32°F. At higher temperatures (41°, 50°, 59°, and 68°F) yellowing, discoloration of stem ends, and decay occurred. Stewart and Barger (1963) noted that moistening cauliflower before vacuum cooling improved cooking quality and significantly maintained moisture content in the head. Ezell and Wilcox (1959 and 1962) found that wilted kale, collard, turnip greens, and cabbage contained less than normal ascorbic acid and carotene. Conditions favorable to wilting, such as high temperature and low relative humidity, hastened the destruction of both ascorbic acid and carotene.

Senescence Inhibitors and Growth Regulators

Green vegetables deteriorate steadily and rapidly after harvest (Zink, 1961). Protein content declines and chlorophyll content decreases; consequently, the loss of quality is inevitable. Kinetin-like chemicals delay senescence of green vegetables and reduce quality loss (Hruschka and Kaufman, 1949). The synthetic analogues of cytokinins have been shown to delay chlorophyll degradation and senescence of leafy vegetables. Dedolph et al. (1963) and MacLean et al. (1963) investigated that pre- and postharvest applications of N^6-benzyladenine reduced the respiration rate and extended the storage life of harvested broccoli. Salunkhe et al. (1962) found that the refrigerated life of cauliflower, radishes, and cabbage was significantly increased by a mild postharvest dip with N^6-benzyladenine and theorized that during the senescence of these vegetables, the primary step in degradation of the soluble ribonucleic acid involved the loss of the end group adenine. A treatment with N^6-benzyladenine, therefore, should provide the necessary adenine and restore the soluble ribonucleic acid molecule. Proteins and amino acids would thus be maintained and the treated produce would stay fresh longer. El Mansy et al. (1967) confirmed the aforestated effectiveness for delaying senescence.

Isbell (1948) reported when kohlrabi and turnip roots were treated and stored with Barsprout (a dust containing methyl ester of naphthalene acetic acid 2.2%) and sprout inhibitor (a dust methyl-l-napthaleneacetate 2.2%), sprouting was delayed. Marth (1952) showed that growth regulators 2-4-D and 2-4-5-T have a pronounced retarding effect on the loss of the green color of broccoli and on cauliflower leaf abscission when applied in the field before or after harvest. However, more concrete information is needed on the toxicity of phenoxy compounds as well as senescence inhibitors to animals.

Microbe Inhibitors

Since antibiotics, fungicides, and hot water are not sterilizing agents, they may inhibit the growth of certain microorganisms on vegetables that are grown in or close to the soil and contain a wide range of both fungi and bacteria. Antibiotics and fungicides of broad spectra are desirable. The nutritional role of antibiotics and fungicides in vegetables at this time appears to be that of increasing the supply of high-quality perishables for a longer time by decreasing their deterioration. Vitamin K and its several analogues, especially K_3 and K_5, have markedly inhibited a number of microorganisms and would add to the nutritive value of vegetables.

Controlled- and Modified-atmosphere Storage

Lebermann et al. (1968) found that in the refrigerated life of broccoli, chlorophyll retention and pH increased progessively with the increase of CO_2 and decrease of O_2. Isenberg and Sayles (1969) reported that the most effective gas combination for storage of Danish cabbage was 2.5% of O_2 and 5% of CO_2 with N_2 making up the difference. With this combination, the weight loss due to respiration was reduced and the cabbage required less trimming. They also noted that with the atmosphere of low O_2 (2.5 to 5%) and high CO_2 (2.5 to 5%) the cabbage retained its green color, succulence, and firmness longer. Lieberman and Hardenburg (1954) showed that when modified atmospheres containing 10, 5, 2.5 or 1% oxygen were flushed over broccoli at 75°F for 3 days, respiration was significantly reduced and yellowing of broccoli was completely inhibited. Isenberg (1973) sprayed cabbage plants one month before harvest with maleic hydrazide, Alar, CCC, N^6-benzyladenine; and Gibberellins subsequently stored them for 7.5 months at 32°F and 96 to 98% R.H. at 5% CO_2 and 1% O_2. Some chemical treatments in combination with the controlled atmosphere reduced weight and trimming losses. In addition, the controlled atmosphere reduced "pepper spot" disease by reducing the rate of senescence. Controlled atmosphere consisting of 2.5% CO_2 and 2.5% O_2 at 32°F with a prestorage treatment of senescence- and microbe-inhibiting chemicals, N^6-benzyladenine and mycostatin, nearly doubled the refrigerated life of polyethylene-packaged coleslaw and cauliflower florets (Salunkhe, 1973).

Radurization

The possibility of increasing the refrigerated life of fresh produce by ionizing radiation is of considerable interest to groups both in and out of the food industry, despite some setbacks. An attractive feature of the process, recognized many years ago, is the negligible temperature rise even at doses that afford complete sterilization. However, radurization (pasteurization with radiation) is not without its undesirable effects, but some progress has been made in eliminating these drawbacks. The promising area, including sprout inhibition in the case of potatoes and onions, can be effectively accomplished with doses ranging from 10 to 15 kilorads, and also the retardation of ripening and microbial growth can be successfully controlled in strawberries, sweet cherries, oranges, and mushrooms with doses of about 0.25 to 0.50 megarad. In general, the success of radurization of perishable fruits and vegetables will depend upon the selection of cultivar and maturity of the commodity, semipermeable packaging material, preirradiation chemical treatment, and storage temperature. Ascorbic acid is the most radiosensitive vitamin. At the pasteurization and sprout inhibition doses, negligible losses of nutrients have taken place (Salunkhe, 1961).

To a great extent, the economic future of the refrigerated fresh produce industry depends upon intelligent use of chemicals — antibiotics, antifungals, senescence inhibitors, and packaging, handling, hydrocooling, radurization, and storage procedures that will extend refrigerated shelf life and maintain quality and wholesomeness. Wider uses of pre- and postharvest treatments that will slow down respiration, delay senescence, and control microbial growth seem inevitable despite increasing legislative limitations on the use of food additives.

Frozen Products

After blanching, there should be a minimum delay in freezing vegetables to avoid microbial changes that can impair flavor, appearance, and nutritional value. When frozen without blanching, significant losses of ascorbic acid occur during storage. The retention of ascorbic acid in frozen products depends upon the water vapor and oxygen permeability of the container.

Eheart (1970) showed that the total ascorbic acid content was higher but pH and chlorophyll were lower in the microwave-blanched than in the water-blanched broccoli. Frozen broccoli packaged in immersible pouches was higher in total acids and somewhat lower in total chlorophyll than that conventionally packaged. Freezer storage caused a decrease in reduced and total ascorbic acid but an increase in acidity and dehydroascorbic acid. Martin et al. (1960) showed that when broccoli was cooked to its optimum doneness, the ascorbic acid retention in frozen as well as fresh samples varied from 65 to 79% depending upon the cooking method. Frozen broccoli showed a slight oxidation of ascorbic acid to dehydroascorbic and diketogulonic acids at 0°F after 36 weeks of storage and no decrease in carotene at 0°F after 61 weeks. Buck and Joslyn (1953, 1956) noted a disturbance in the normal glycolytic reaction and the formation of off-flavors in broccoli when scalded and frozen. These off-flavors were caused by the formation and accumulation of ethyl alcohol and acetaldehyde. These authors further noted that aldehydic and ketonic compounds might serve as precursors of the off-flavors developing in under-scalded frozen broccoli.

Dietrich et al. (1962) showed that the reduction in quality and ascorbic acid of frozen cauliflower was fourfold for each 10°F rise in temperature when stored at from −10 to 40°F. Fluctuating temperatures produce irreversible and cumulative changes. Dietrich et al. (1957) noted that a summation of reduced ascorbic acid and its oxidative products (dehydroascorbic and diketogulonic acids) provided a more reliable indicator in the stored vegetables than did the reduced ascorbic acid content. The accumulation of diketogulonic acid as storage progresses seems to cause parallel adverse changes during storage. In addition, the conversion of chlorophyll to pheophytin is correlated with storage temperatures.

Van den Berg (1961) noted that the magnitude of the change in pH of the stored products appeared to be controlled by the buffer capacity of proteins and soluble salts. Fitzgerald and Fellers (1938) showed that frozen cruciferous vegetables in particular retained their vitamin A; and storage at a low temperature in an impervious package out of contact with ultraviolet light was conducive to complete retention of vitamin A up to the time of delivery to the consumer. Zscheile et al. (1943) showed that blanching before quick freezing retarded the loss of carotene during storage. Eheart (1969) showed that the retention of ascorbic acid, acidity, and chlorophyll depended upon the cultivar of broccoli. The boil-in-bags retained significantly higher amounts of ascorbic acid and chlorophyll than freshly cooked broccoli. Four-months freezer-stored blanched samples had an increase in acidity and a loss of chlorophyll and ascorbic acid. Cooking the stored samples caused a greater loss in chlorophyll than ascorbic acid.

The loss of moisture from frozen vegetables during the storage is associated with water vapor transfer properties of the packaging material. The degree of carotene and ascorbic acid losses varies from vegetable to vegetable and with storage temperature. It is difficult to generalize in regard to the retention of ascorbic acid content. Each vegetable should be considered individually with respect to blanching, packaging, time, and temperature of storage. The B vitamins appear to be retained to a high degree during storage. Data concerning the influence of commercial storage conditions on the retention of these nutrients are extremely limited.

Canned Products

The influence of storage on the amount and quality of protein in terms of essential amino acids in canned foods has received little attention. Nonenzymatic browning reactions accompany storage. The extent of such reactions depends upon the concentrations of reducing sugars and amino acids. Head space oxygen is important to the stability of ascorbic acid in acid foods packaged either in glass or in plain tin cans. With less acid foods such as certain vegetables, there was less influence on the ascorbic acid content.

Among the nutrients, next to ascorbic acid, thiamine is probably most affected by the storage temperature. In the range of 50° to 80°F, increases in temperature were found to have greater influence on thiamine retention in less-acid vegetables than in more-acidic fruits. Farrer (1955) and Labuza (1973) discussed the kinetical approach to analyzing nutrient retention data in processed foods when stored at different temperatures. By this approach, it is possible to quickly compare results from different laboratories to predict the retention of nutrients in processed foods stored under different time-temperature conditions.

Fermented Products

Cabbage is the only cruciferous vegetable preserved in large quantities by fermentation. Occasionally cauliflower is mixed with other vegetables for pickling. In the United States, sauerkraut is one of the most important fermented vegetable products. The quantity is exceeded only by cucumber pickles. Sauerkraut production in the United States is nearly 200,000 tons. In recent years, however, the production has steadily declined. Nabors and Salunkhe (1969) showed the importance of *Leuconostoc mesenteroides* and *Lactobacillus plantarum* on the quality of fresh and dehydrated sauerkraut. The quality and nutritive value of sauerkraut depend upon temperature, sodium chloride, and microflora during fermentation. Sauerkraut and its juice contain a plentiful supply of ascorbic acid. Tarkhov and Martishenya (1933) and later Martishenya and Tarkhov (1936) observed the loss of ascorbic acid during the long storage of sauerkraut.

Pederson et al. (1939) and Pederson and Robinson (1952) noted that sauerkraut contains as much ascorbic acid as the original cabbage. However, sauerkraut stored at high temperature loses its ascorbic acid content and deteriorates rapidly as shown by the concomitant changes in color, flavor, and texture. The deterioration was less in tin cans than in glass containers. The latter losses may be due to light intensity. Pederson (1956) found that the retention of ascorbic acid in sauerkraut depended upon the fermentation temperature and length of time in the filling vat. Similarly, he concluded that the storage time and temperature also influence the ascorbic acid retention in canned sauerkraut. If more attention were given to these findings, a product with high ascorbic acid content could be obtained more constantly. Sedky et al. (1953) reported that discoloration and off-flavor of sauerkraut were due to the degradation of ascorbic acid content when exposed to light.

Dehydrated Products

Oxidation is the primary cause of losses, particularly of ascorbic acid, during drying and dehydration. This can be reduced by the application of a vacuum or by flushing the product with nitrogen gas and, if the drying time is shortened, without necessitating a marked change in temperature. Vitamin C also may be converted to dehydroascorbic acid in a nonenzymatic manner following its oxidation. Morgan et al. (1945) indicated that dehydrated broccoli lost 40% of its thiamine during storage but that the losses of niacin, riboflavin, and pantothenic acid were insignificant. Moyer (1943) studied the carotene, thiamine, and ascorbic acid contents of cabbage and rutabaga prior to and during dehydration, after storage, and after cooking. Though there was little loss of carotene during cooking, its loss was rapid at all storage temperatures. Storage under carbon dioxide tended to diminish the loss. Some of the thiamine was leached out during blanching and prior to dehydration and very little original ascorbic acid was retained after dehydration. Samples stored below 33°F were unaffected. At high storage temperatures, the color faded. Using large amounts of water during the cooking of dehydrated vegetables brought about a decrease in carotene, thiamine, and ascorbic acid. Vitamin stability, particularly of ascorbic acid and carotene, however, can be increased by reducing the moisture. Craft (1944) showed that during the steam blanching of cabbage, fleshy parenchyma cells and intercellular spaces must be free from air; otherwise, during dehydration, the product would turn brown. Reeve (1943) reemphasized the disadvantage of oil being present in vegetables prepared for commercial dehydration. The oils may be involved in the oxidation and subsequent "off"-flavor formation during dehydration and storage of the products. The oil may be in the form of lipoproteins and/or lecithin. Reeve further substantiated the beneficial effects of blanching in reducing vitamin losses due to the inactivation of enzymes. Eheart and Sholes (1946) showed that a 30-sec soaking treatment in 1% Na_2SO_3 solution was the best method to preserve ascorbic acid in dehydrated cabbage.

Ranganna and Setty (1968) reported that dehydroascorbic acid and 2,3-diketogulonic acid, formed from ascorbic acid during the dehydration of cabbage, react nonenzymatically with the free amino acids to cause red to brown discoloration in the dried product. The conversion of a red color to brown was accompanied by the liberation of the carbonyl group. Harris (1945) noted a con-

siderable loss of carotene and ascorbic acid in broccoli and Brussels sprouts when dehydrated and preserved with sodium chloride.

In-package Desiccant

The use of an in-package desiccant to bring the moisture content to 1% or lower should permit the storage of dehydrated vegetables for six months or more at 70 to 100°F without significant losses of vitamins.

Compression of Dehydrated Products

Compression might reduce oxidative degradation of the dehydrated vegetables since air is removed during compression. Thus the retention of ascorbic acid and carotene could be increased.

Packaged Products

Many of the losses in refrigerated, frozen, canned, and dehydrated products can be reduced by the proper selection and use of packaging materials and methods. Correct packaging can control losses from the effects of light, oxygen, moisture, temperature, and microorganisms. Interactions between food and packaging material, however, may cause a degradation of nutrients in food. Light can cause degradation of light-sensitive vitamins including riboflavin, carotene, thiamine, and amino acids in foods packaged in transparent containers. A high storage temperature accelerates microbial growth and degradation of many nutrients such as ascorbic acid, thiamine, and amino acids in packaged products.

Home and Institutional Cooking Processes

Cooking for home or institutional feeding is the final link in food production, transportation, processing, and storage. Nutrient losses that occur when foods are cooked depend upon the freshness of the food, size of the product, type of equipment, amount of water for cooking, temperature of cooking, length of cooking time, and the amount of salt in the cooking water.

Krehl and Winters (1950) reported that while over 20% of the calcium was extracted from cabbage during the boiling, only 9% was extracted during pressure cooking. Nutrient losses are caused in part by extraction and in part by destruction. The retention of riboflavin in 400 g of cabbage when 200 and 800 ml of water was used was 74 and 50%, respectively (Van Duyne et al., 1948), and the loss of ascorbic acid was halved when the cooking pan was covered (Noble and Waddell, 1945). The length of cooking time is important. The retention of ascorbic acid was 64%, 57%, and 55% when broccoli was boiled for 2, 5.5, and 11 min, respectively (Barnes et al., 1943a). Broccoli had 79% folic acid retention (Causey and Fenton, 1951) subsequent to boiling.

Munsell et al. (1949) reported retention of 67% ascorbic acid, 88% thiamine, and 100% riboflavin in steamed cabbage as compared with the retention of 30% ascorbic acid, 43% thiamine, and 50% riboflavin in boiled cabbage. McIntosh et al. (1940) compared the retention of ascorbic acid in vegetables when cooked by boiling, by steaming, and by pressure to the identical doneness. The retention by the three methods of cooking Brussels sprouts was 77%, 91%, and 97% and the cauliflower 81%, 77%, and 92%, respectively. Thus the best results were obtained by pressure cooking. Slow cooking usually results in greater losses, especially in an uncovered pan. Losses were two to six times greater when vegetables were boiled in a copper pan than in a glass one and the losses were lower when the amount of cooking water was restricted. Dietrich and Neumann (1965) found that blanching Brussels sprouts with water inactivated enzymes faster and degraded chlorophyll less than blanching with steam. The latter gave a dull-colored and soft product. At 190°, 200°, and 212°F, the ascorbic acid content was 0 to 10% higher in the steam-blanched than in the water-blanched samples. Storage at 20°F for 4 to 8 weeks caused considerable deterioration of chlorophyll and ascorbic acid.

Most of the nutrients lost from vegetables during cooking are leached into the cooking water. Broccoli boiled for 2, 5.5, and 11 min lost 25%, 32%, and 33% of its ascorbic acid, respectively (Barnes et al., 1943a). These losses can be minimized by reducing the volume of cooking water. Barnes et al. (1943a) noted an ascorbic acid retention of 82%, 57%, and 53% when broccoli was cooked in 100 ml, 500 ml, and 1,000 ml of water, respectively. McIntosh et al. (1942) observed retentions of 84%, 63%, and 65% ascorbic acid content when cauliflower was cooked in 120 ml, 250 ml, and 480 ml of water. According to Van Duyne et al. (1948), when 400 g of cabbage was cooked in 200 ml and 800 ml of water, the riboflavin retention was 74% and 50%, respectively.

Fenton and Gifft (1943) suggested that frozen

vegetables should not be thawe[d] cooking but should be place[d] minimum quantity of rapidly b[oiling] and Odland (1973) reported NH_4OH and NH_4HCO_3 preve[nt] losses during the blanching and .... of green beans. A trained panel rated NH_4HCO_3 treated beans higher than controls in the acceptability of color, texture, and flavor. Munsell et al. (1949) measured the nutrients in the cooking water recovered from cabbage and found that it contained more thiamine, riboflavin, and niacin than the cabbage itself. Essentially all of the thiamine and riboflavin had been extracted but none was destroyed. Sodium bicarbonate and other alkaline salts are occasionally added to maintain the green color and to increase the rate and temperature of the cooking. This practice is destructive to water-soluble nutrients, especially thiamine and ascorbic acid, that are sensitive to high temperature and alkali.

Microwave Cooking

Thomas et al. (1949) showed that the cooking time and amount of water used were determining factors in retention of nutrients in vegetables. The pressure cooking technique was the best in the retention of nutrients as compared to the boiling water or microwave technique. Van Zante and Johnson (1970) stated that thiamine and riboflavin in buffered aqueous solutions heated with microwave and conventional electric devices exhibited the greatest retention of both vitamins when they were in higher concentrations. They also noted that no significant difference in vitamin contents resulted from the two different heating methods. Eheart and Gott (1964) noted no significant difference in the retention of ascorbic acid in broccoli when cooked by microwave or conventional methods. Kylen et al. (1961) reported that shortening either microwave or conventional cooking time had no significant effect on the retention of ascorbic acid.

Gordon and Noble (1959) found that cauliflower and cabbage, when cooked by microwave and by pressure saucepan, were similar in flavor but broccoli cooked by microwave was milder in flavor than those cooked in the pressure saucepan. More ascorbic acid was retained by vegetables cooked by microwave than by those cooked in the pressure saucepan and the difference was greater relative to those cooked in boiling water. Campbell

_1958) noted that the ascorbic acid contents _abbage and broccoli cooked with microwaves _ere significantly higher than in those cooked with conventional heating. Additional research is needed before microwave cooking becomes a standard method of cooking in homes and restaurants.

By the use of small amounts of water or of steam, nutrient leaching can be minimized; and by shortening the cooking time through the use of a stainless-steel pressure cooker, the extraction and destruction of nutrients can be reduced significantly. Since cooking water contains significant quantities of starch, vitamins, minerals, and amino acids, it should not be discarded but should be used in soups or gravy.

The effects of cooking upon the ascorbic acid content of foods have received the most attention, while fewer studies have been conducted on thiamine, even less on riboflavin, niacin, and carotene, with practically no work on other vitamins, amino acids, and minerals.

NEEDS FOR FUTURE RESEARCH

The future of the cruciferous vegetable industry in this country and elsewhere will depend upon the stimulation of greater public awareness of the value of these highly-nutritious and low-calorie foods. Cruciferous growers, processors, and shippers must promote and advertise these commodities as Sunkist growers and other processors have done for citrus products. Consumer education through industry, U.S. Department of Health, Education, and Welfare, U.S. Department of Agriculture, and universities through their extension divisions should carry the message of better nutrition from crucifers via 4-H and FHA clubs, home demonstration councils, and vocational schools and publicity through cooking demonstrations by appliance manufacturers and electricity and gas companies.

Basic research is needed on the (1) interactions of amino acids, vitamins, and minerals and their retention and availability; (2) on the stability of pigments and flavor compounds of processed and stored cruciferous products; (3) on maintaining a firm texture of the processed products through a study of histology, histochemistry, and electron microscopy of the cells; (4) on how to retain cell integrity during processing; (5) on biological data including digestibility and availability of the major nutrients present in crucifers; and (6) on developing new cultivars with lower concentrations of sulfur compounds for mild-flavored products.

Studies need to be conducted on improving the nutritive value by increasing plant efficiency by preplanting and preharvesting applications of _s_-triazine and 1,3-dichloropropene compounds which have been proven to be able to significantly increase the mineral, vitamin, protein, and amino acid contents in leaves, buds, flowers, stems, and roots (Salunkhe et al. 1971b, c, and Singh et al. 1972).

Research is needed on controlled-, modified-, and hypobaric-atmosphere storage in conjunction with hydrocooling and senescence and microbe inhibiting chemicals to extend the shelf life of fresh and refrigerated produce while retaining nutrients for a longer time. Nutritional stability research in this area is meager.

Research on the stability of vitamins, particularly pantothenic acid, folacin, and vitamin B_{12}, in cruciferous vegetables is lacking.

More practical research is needed on: (1) the feasibility of fortification with dehydrated and powdered cruciferous leaves in cereal meal to supply certain amino acids (lysine, methionine, tryptophan, and cystine), vitamins (C, B complex, and carotene), and minerals (sulfur, calcium, iron, and magnesium), and to improve flavor, color, and wholesomeness; (2) "water-less" cooking with steam and microwaves and prolonged heating with infrared on the nutritional value of crucifers; (3) new recipes with simple snappy ideas such as "shake and bake" and "roast and boast"; (4) preparations to improve quality by reducing sulfur by a brief blanching with weak acids and masking flavor with certain potentiator(s), sauce(s) and spice(s) and subsequently packaging them in boilable pouches to minimize nutritional losses during cooking; (5) adaptation or modification of the Chinese method of sautéing vegetables to maintain firm texture and crispness without significant loss of color and nutritive value by using high temperature for a short time in a minimal of water. In addition, an evaluation should be made on cooking methods from other cultures such as "Kosher," "Slavic," "Italian," "Spanish," and "Soul" which use substantial quantities of cruciferous vegetables in the diet.

ACKNOWLEDGMENTS

The authors wish to thank Dr. R. G. Garner and Dr. Marie McCarthy for furnishing several USDA publications which were most useful in the preparation of this paper and Professor Willard B. Robinson, Dr. Alice Denney, and Mr. Ted Wilde for their critical reviews of the manuscript.

REFERENCES

Aberg, B. and Ekdahl, I., Effect of nitrogen fertilization on the ascorbic acid content of green plants, *Physiol. Plant.,* 1, 290, 1948.

Agricultural Research Service, Folic Acid in Foods, Agriculture Handbook No. 29, U. S. Department of Agriculture, Washington, D. C., 1951.

Agricultural Research Service, Composition of Foods, Agriculture Handbook No. 8, U. S. Department of Agriculture. Washington, D. C., 1963.

Agricultural Research Service, Pantothenic Acid, Vitamin B_6, and Vitamin B_{12} in Foods, Home Economics Research Report No. 36, U. S. Department of Agriculture, Washington, D. C., 1969.

Agricultural Research Service, Food and Nutrient Intake of Individuals in the United States, Spring 1965, Household Food Consumption Survey 1965-66, Report No. II, U. S. Department of Agriculture, Washington, D. C., 1972.

Bailey, S. D., Bazinet, M. L., Driscoll, J. L., and McCarthy, A. I., The volatile sulfur components of cabbage, *J. Food Sci.,* 26, 163, 1961.

Barnes, B., Tressler, D. K., and Fenton, F., Effect of different cooking methods on the vitamin C content of quick-frozen broccoli, *Food Res.,* 8, 13, 1943a.

Barnes, B., Tressler, D. K., and Fenton, F., Thiamine content of fresh and frozen peas and corn before and after cooking, *Food Res.,* 8, 420, 1943b.

Branion, H. D., Roberts, J. S., Cameron, C. R., and McCready, A. M., The ascorbic acid content of cabbage, *J. Am. Diet. Assoc.,* 24, 101, 1948.

Buck, P. A. and Joslyn, M. A., Accumulation of alcohol in underscalded frozen broccoli, *J. Agric. Food Chem.,* 1, 309, 1953.

Buck, P. A. and Joslyn, M. A., Formation of alcohol, acetaldehyde, and acetone in frozen broccoli tissue, *J. Agric. Food Chem.,* 4, 548, 1956.

Burrell, R. C., Brown, H. D., and Ebright, V. R., Ascorbic acid content of cabbage as influenced by variety, season, and soil fertility, *Food Res.,* 5, 247, 1940.

Burton, B. T., *The Heinz Handbook of Nutrition,* McGraw-Hill Book Company, Inc., New York, 1959.

Calloway, D. H. and Munson, A. H., Response of cereal-fed guinea pigs to dietary broccoli supplementation and X-irradiation, *J. Nutr.,* 73, 191, 1961.

Calloway, D. H., Newell, G. W., Calhoun, W. K., and Munson, A. H., Further studies of the influence of diet on radio-sensitivity of guinea pigs, with special reference to broccoli and alfalfa, *J. Nutr.,* 79, 340, 1963.

Campbell, C. L., Lin, T. Y., and Proctor, B. E., Microwave vs. conventional cooking. I. Reduced and total ascorbic acid in vegetables, *J. Am. Diet. Assoc.,* 34, 365, 1958.

Causey, K. and Fenton, F., Effect of reheating on palatability, nutritive value, and bacterial count of frozen cooked foods I. Vegetables, *J. Am. Diet. Assoc.,* 27, 390, 1951.

Cheney, G., Cinchophen gastric ulcers in chicks, *Am. Med. Assoc. Arch. Intern. Med.,* 70, 532, 1942.

Cheney, G., Rapid healing of peptic ulcers in patients receiving fresh cabbage juice, *Calif. Med.,* 70(1), 10, 1949.

Cheney, G., Antipeptic ulcer dietary factor (vitamin U) in the treatment of peptic ulcer, *J. Am. Diet. Assoc.,* 26, 668, 1950.

Cheney, G., Vitamin U therapy of peptic ulcer, *Calif. Med.,* 77, 248, 1952.

Cheney, G., Vitamin U concentrate therapy of peptic ulcer, *Am. J. Gastroenterol.,* 21, 230, 1954.

Cheney, G., The medical arrangement of gastric ulcers with vitamin U therapy, *Stanford Med. Bull.,* 13, 204, 1955.

Cheney, G., Waxler, S. H., and Miller, I. J., Vitamin U therapy of peptic ulcer, *Calif. Med.,* 84(1), 39, 1956.

Chiang, J. C., Singh, B., and Salunkhe, D. K., Effect of water quality on canned carrots, sweet cherries, and apricots, *J. Am. Soc. Hort. Sci.,* 96, 353, 1971.

Clapp, R. C., Long, L., Dateo, G. P., Bissett, F. H., and Hasselstrom, T., The volatile isothiocyanates in fresh cabbage, *Am. Chem. Soc. J.,* 81, 6278, 1959.

Clayton, M. M. and Borden, R. A., The availability for human nutrition of the vitamin C in raw cabbage and home-canned tomato juice, *J. Nutr.,* 25, 349, 1943.

Clayton, M. M. and Goos, C., Effect of French dressings, vinegars, and acetic acid on rate of loss of vitamin C in raw cabbage, *Food Res.,* 12, 27, 1947.

Craft, A. S., Cellular changes in certain foods and vegetables during blanching and dehydration, *Food Res.,* 9, 442, 1944.

Crosby, D. G., Natural cholinesterase inhibitors in food, in *Toxicants Occurring Naturally in Foods,* Publication No. 1354, National Academy of Science, National Research Council, Washington, D. C., 1966, 112.

Dateo, G. P., Clapp, R. C., MacKay, D. A. M., Hewitt, E. J., and Hasselstrom, T. Identification of the volatile sulfur components of cooked cabbage and the nature of precursors in the fresh vegetable, *Food Res.,* 22, 440, 1957.

Dedolph, R. R., Wittwer, S. H., and Larzelere, H. E., Consumer verification of quality maintenance induced by N^6-benzyladenine in the storage of celery (*Apium graveolens*) and broccoli (*Brassica oleracea* var. *Italica*), *Food Technol.* 17, 1323, 1963.

Del Valle, C. G., and Harmon, S. A., Collard growth and mineral composition as influenced by soil temperature and two fertility levels, *Proc. Am. Soc. Hort. Sci.,* 91, 347, 1967.

Dietrich, W. C., Lindquist, F. E., Miers, J. C., Bohart, G. S. Neumann, H. J., and Talburt, W. F. The time-temperature tolerence of frozen foods. IV. Objective tests to measure adverse changes in frozen vegetables, *Food Technol.* 11, 109, 1957.

Dietrich W. C., Nutting, M. D. F., Boggs. M. M., and Weinstein, N. E., Time-temperature tolerance of frozen foods. XXIV. Quality changes in cauliflower, *Food Technol.,* 16, 123, 1962.

Dietrich, W. C. and Neumann, H. J. Blanching of Brussels sprouts, *Food Technol.,* 19, 1174, 1965.

Dietrich, W. C., Huxsoll, C. C., and Gudagani, D. G. Comparison of microwave, conventional, and combination blanching of Brussels sprouts for frozen storage, *Food Technol.,* 24, 613, 1970.

Economic Research Service, National Food Situation, NFS-138, U.S. Department of Agriculture, Washington, D.C., 1971.

Eheart, J. F., Young, R. W., Massey, P. H., Jr., and Havis, J. R., Crop, light intensity, soil pH, and minor element effects on the yield and vitamin content of turnip greens, *Food Res.,* 20, 575, 1955.

Eheart, M. S. and Scholes, M. L., Effects of methods of blanching and temperature storage on nutritive value of dehydrated cabbage, *Food Res.,* 11, 298, 1946.

Eheart, M. S. and Scholes, M. L. Effects of old-fashion and modern methods of cooking on the retention of nutrients in vegetables. I. Cabbage: influence of methods of reporting data on results, *Food Res.,* 13, 106, 1948.

Eheart, M. S. and Gott, C. Conventional and microwave cooking of vegetables, *J. Am. Diet. Assoc.,* 44, 116, 1964.

Eheart, M. S., Fertilization effects on the chlorophylls, carotene, pH, total acidity, and ascorbic acid in broccoli, *J. Agric. Food Chem.,* 14, 18, 1966.

Eheart, M. S., Effect of microwave vs. water-blanching on nutrients in broccoli, *J. Am. Diet. Assoc.,* 50, 207, 1967.

Eheart, M. S., Variety, fresh storage, blanching solution, and packaging effects on ascorbic acid, total acids, pH, and chlorophylls in broccoli, *Food Technol.,* 23, 238, 1969.

Eheart, M. S., Effect of storage and other variables on composition of frozen broccoli, *Food Technol.,* 24, 1009, 1970.

Eheart, M. S. and Odland D., Storage of fresh broccoli and green beans, *J. Am. Diet. Assoc.,* 60, 402, 1972.

Eheart, M. S. and Odland D., Use of ammonium compounds for chlorophyll retention in frozen green vegetables, *J. Food Sci.,* 38, 202, 1973.

Einstein, M. A. and Hornstein, I., Food preference of college students and nutritional implications, *J. Food Sci.,* 35, 429, 1970.

El Mansy, H. I., Salunkhe, D. K., Hurst, R. L., and Walker, D. R., Effect of pre- and post-harvest applications of 6-furfurylaminopurine and N^6-benzyladenine on physiological and chemical changes in lettuce, *Hort. Res.,* 7, 81, 1967.

Ezell, B. D. and Wilcox, M. S., Loss of vitamin C in fresh vegetables as related to wilting and temperature, *J. Agric. Food Chem.,* 7, 507, 1959.

Ezell, B. D. and Wilcox, M. S., Loss of carotene in fresh vegetables as related to wilting and temperature, *J. Agric. Food Chem.,* 10, 124, 1962.

Fager, E. E. C., Olson, O. E., Burris, R. H., and Elvehjem, C. A., Folic acid in vegetables and certain other plant materials, *Food Res.,* 14, 1, 1949.

Fariis, P. and Kjaer, A., 4-Methylthio-3-butenyl isothiocyanate, the pungent principle of radish root, *Acta Chem. Scand.,* 20, 698, 1966.

Farrer, K. T. H., The thermal destruction of vitamin B in foods, in *Advances in Food Research,* Vol. 6, Stewart, G. F., Ed., Academic Press, Inc., New York, 1955, 257.

Fenton, F. and Gifft, H., Palatability studies of commercially dehydrated vegetables, *Food Res.,* 8, 364, 1943.

Fitzgerald, G. A. and Fellers, C. R., Carotene and ascorbic acid content of fresh market and commercially frozen fruits and vegetables, *Food Res.,* 3, 109, 1938.

Food and Nutrition Board *Recommended Dietary Allowances,* 7th ed., National Research Council, National Academy of Sciences, Washington, D. C., 1968.

Food Policy and Food Science Service, Amino-Acid Content of Foods and Biological Data on Proteins, Food and Agriculture Organization of the United Nations, Rome, Italy, 1970.

Foter, M. J. and Golick, A. M., Inhibitory properties of horseradish vapors, *Food Res.,* 3, 609, 1938.

Foter, M. J., Bactericidal properties of allyl isothiocyanate and related oils, *Food Res.,* 5, 147, 1940.

Francis, F. J., Discoloration and quality maintenance in cole slaw, *Proc. Am Soc. Hort. Sci.,* 75, 449, 1960.

Gordon, J. and Noble, I., Comparison of electronic vs. conventional cooking of vegetables, *J. Am. Diet. Assoc.,* 35, 241, 1959.

Greer, M. A., Isolation from rutabaga seed of progoitrin, the precursor of the naturally occurring antithyroid compound, goitrin (1-5-vinyl-2-thiooxazolidone), *Am. Chem. Soc. J.,* 78, 1260, 1956.

Greer, M. A., The natural occurrence of goitrogenic agents, *Recent Progr. Horm. Res.,* 18, 187, 1962.

Gustafson, F. G., Influence of temperature on the vitamin content of green plants, *Plant Physiol,* 25, 150, 1950.

Haller, M. H., Effect of root-trimming, washing, and waxing on the storage of turnips, *Proc. Am. Soc, Hort. Sci.,* 50, 325, 1947.

Hamner, K. C. and Parks, R. Q., Effect of light intensity on ascorbic acid of turnip greens, *J. Am. Soc. Agron.,* 36, 269, 1944.

Hansen, E., Seasonal variations in the mineral and vitamin content of certain green vegetable crops, *Proc. Am. Soc. Hort. Sci.,* 46, 299, 1945a.

Hansen, E., Variations in the carotene content of carrots. *Proc. Am. Soc. Hort. Sci.,* 46, 355, 1945b.

Harris, S. C., Carotene and ascorbic acid in fresh and salted vegetables, *J. Am. Diet. Assoc.,* 21, 360, 1945.

Harris, R. S. and Von Loesecke, H., *Nutritional Evaluation of Food Processing,* 2nd Printing, Avi Publishing Co., Westport, Conn., 1971.

Hewitt, E. J., MacKay, D. A., Konigsbacher, K., and Hasselstrom, T., The role of enzymes in food flavors, *Food Technol.,* 10, 487, 1956.

Hing, F. S. and Weckel, K. G., Some volatile components of cooked rutabaga, *J. Food Sci.,* 29, 149, 1964.

Hirsch, J. A., Niles, A. D., and Kemmerer, A. R., The essential amino acid content of several vegetables, *Food Res.,* 17, 442, 1952.

Holley, R. W., Boyle, F. P., and Durfee, H. K., A study of the auxins in cabbage using counter-current distribution, *Arch. Biochem. Biophys.,* 32, 192, 1951.

Holmes, A. D., Crowley, L. V., and Kuzmeski, J. W., Influence of supplementary calcium and magnesium fertilizers upon nutritive value of kale, *Food Res.,* 10, 401, 1945.

Howard, F. D., MacGillivray, J. H., and Yamaguchi, M., Nutrient composition of fresh California-grown vegetables, California Agricultural Experiment Station Bulletin 788, Davis, Calif., 1962.

Hruschka, H. W. and Kaufman, J., Storage tests with Long Island cauliflower to inhibit leaf abscission by using plant growth regulators, *Proc. Am. Soc. Hort. Sci.,* 54, 438, 1949.

Isbell, C. L., Effect of sprout preventive treatments on fall-harvested kohlrabi, potatoes, sweet potatoes, and turnips while in storage, *Proc. Am. Soc. Hort. Sci.,* 52, 368, 1948.

Isenberg, F. M. and Sayles, R. M., Modified atmosphere storage of Danish cabbage, *J. Am. Soc. Hort. Sci.* 94, 447, 1969.

Isenberg, F. M., personal communication, 1973.

Janes, B. E., The relative effect of variety and environment in determining the variations of per cent dry weight, ascorbic acid, and carotene content of cabbage and beans, *Proc. Am. Soc. Hort. Sci.,* 45, 387, 1944.

Jensen, K. A., Conti, J., and Kjaer, A., Isothiocyanates. II. Volatile isothiocyanate in seeds and roots of various *Brassicae, Acta Chem. Scand.,* 7, 1267, 1953.

Johnson, A. E., Nursten, H. E., and Williams, A. A., Vegetable volatiles: A survey of components identified — Part I, *Chem. Ind. (Lond.),* p. 556, 1971a.

Johnson, A. E., Nursten, H. E., and Williams, A. A., Vegetable volatiles: A survey of components identified — Part II, *Chem. Ind. (Lond.),* p. 1212, 1971b.

Kamen, J. M., Peryam, D. R., Peryam, D. B., and Kroll, B. J., 1963 Analysis of U. S. Army Food Preference Survey Report No. 67, Department of Army, Washington, D. C., 1967.

Kelly, J., The availability of the calcium of some New Zealand vegetables, *J. Nutr.* 25, 303, 1943.

Kemmerer, A. R. and Acosta, R., The essential amino acid content of several vegetables, *J. Nutr.,* 38, 527, 1949.

Kjaer, A., Naturally derived isothiocyanates (mustard oils) and their parent glucosides, in *The Chemistry of Organic Natural Products,* Zechmeister, L., Ed., Springer-Verlag. Vienna, 1960, 122.

Krehl, W. A. and Winters, R. W., Effect of cooking methods on retention of vitamins and minerals in vegetables, *J. Am. Diet. Assoc.,* 26, 966, 1950.

Kylen, A. M., Charles, V. R., McGrath, B. H., Schleter, J. M., West, L. C., and Van Duyne, F. O., Microwave cooking of vegetables, *J. Am. Diet. Assoc.,* 39, 321, 1961.

Labuza, T. P., Effects of dehydration and storage (symposium: effects of processing, storage, and handling on nutrient retention in foods), *Food Technol.,* 27(3), 20, 51, 1973.

Lakhanpal, R. K., Davis, J. R., Typpo, J. T., and Briggs, G. M., Evidence for an unidentified growth factor(s) from alfalfa and other plant sources for young guinea pigs, *J. Nutr.,* 89, 341, 1966.

Lebermann, K. W., Nelson, A. I., and Steinberg, M. P., Post-harvest changes of broccoli stored in modified atmospheres. II. Acidity and its influence on texture and chlorophyll retention of the stalks, *Food Technol.,* 22, 490, 1968.

Levine, H., Culp, F. B., and Anderson, C. B., The value of some vegetables in nutritional anemia, *J. Nutr.,* 5, 295, 1932.

Lichtenstein, E. P., Strong, F. M., and Morgan, D. G., Identification of 2-phenylethylisothiocyanate as an insecticide occurring naturally in the edible part of turnips, *J. Agric. Food Chem.,* 10, 30, 1962.

Lichtenstein, E. P., Morgan, D. G., and Mueller, C. H., Naturally occurring insecticides in cruciferous crops, *J. Agric. Food Chem.,* 12, 158, 1964.

Lieberman, M. and Hardenburg, R. E., Effect of modified atmospheres on respiration and yellowing of broccoli at 75 degree F, *Proc. Am. Soc. Hort. Sci.,* 63, 409, 1954.

Lyons, J. M. and Rappaport, L., Effect of temperature on respiration and quality of Brussels sprouts during storage, *Proc. Am. Soc. Hort. Sci.,* 73, 361, 1959.

MacLean, D. C., Dedolph, R. R., and Wittwer, S. H., Respiratory responses of broccoli to pre- and post-harvest treatments with N^6-benzyladenine, *Proc. Am. Soc. Hort. Sci.,* 83, 484, 1963.

MacLeod, A. J. and MacLeod, G., The flavor volatiles of dehydrated cabbage, *J. Food Sci.,* 35, 739, 1970a.

MacLeod, A. J. and MacLeod, G., Effects of variations in cooking methods on the flavor volatiles of cabbage, *J. Food, Sci.,* 35, 744, 1970b.

Mallette, M. F., Dawson, C. R., Nelson, W. L., and Gortner, W. A., Commercially dehydrated vegetables. Oxidative enzymes, vitamin content, and other factors, *Ind. Eng. Chem.,* 38, 437, 1946.

Marth, P. C., Effect of growth regulators on the retention of color in green sprouting broccoli, *Proc. Am. Soc. Hort. Sci.,* 60, 367, 1952.

Martin, M. E., Sweeney, J. P., Gilpin, G. L., and Chapman, V. J., Factors affecting the ascorbic acid and carotene content of broccoli, *J. Agric. Food Chem.,* 8, 387, 1960. ✗

Martishenya, A. I. and Tarkhov, V. S., Sauerkraut as a source of antiscorbutic vitamin. *Chem. Zentralblatt,* 11:1197, 1936. 32, 3847, 1938.

Maruyama, F. T., Identification of dimethyl trisulfide as a major aroma component of cooked *Brassicaceous* vegetables, *J. Food Sci.,* 35, 540, 1970.

Maynard, D. N., Gersten, B., and Vernell, H. F., The distribution of calcium as related to internal tipburn, variety, and calcium nutrition in cabbage, *Proc. Am. Soc. Hort. Sci.,* 86, 392, 1964.

McIntosh, J. A., Tressler, D. K., and Fenton, F., The effect of different cooking methods on the vitamin C content of quick-frozen vegetables, *J. Home Econ.,* 32, 692, 1940.

McIntosh, J. A., Tressler, D. K., and Fenton, F., Ascorbic acid content of five quick-frozen vegetables, *J. Home Econ.,* 34, 314, 1942.

McRorie, R. A., Sutherland, G. L., Lewis, M. S., Barton, A. D., Glazeuer, M. R., and Shive, W., Isolation and identification of a naturally occurring analog of methionine, *Am. Chem. Soc. J.,* 76, 115, 1954.

Morgan, A. F., Mackinney, G., and Cailleau, R., Losses of ascorbic acid and four B vitamins in vegetables as a result of dehydration, storage, and cooking, *Food Res.,* 10, 5, 1945. ✗

Moyer, J. C., The nutritive value of dehydrated vegetables, *J. Am. Diet. Assoc.,* 19, 13, 1943.

Munsell, H. E., Streightoff, F., Bendor, B., Orr, M. L., Ezekiel, S. R., Leonard, M. H., Richardson, M. E., and Koch, F. G., Effect of large-scale methods of preparation on the vitamin content of food. III. Cabbage, *J. Am. Diet. Assoc.,* 25, 420, 1949.

Murphy, E. L., personal communication, 1973.

Nabb, D. P. and O'Dell, B. L., Influence of dietary factors upon *Salmonella typhimurium* infection in the guinea pig, *J. Nutr.,* 84, 191, 1964.

Noble, I. and Waddell, E., Effect of different methods of cooking on ascorbic acid content of cabbage, *Food Res.,* 10, 246, 1945.

Nabors, W. T. and Salunkhe, D. K., Pre-fermentation inoculations with *Leuconostoc mesenteroides* and *Lactobacillus plantarum* on physicochemical properties of fresh and dehydrated sauerkraut, *Food Technol.,* 23(3), 67, 1969.

Nylund, R. E., Ascorbic acid content of twenty-five varieties of the rutabaga *(Brassica napobrassica),* *Proc. Am. Soc. Hort. Sci.,* 54, 367, 1949.

Oberg, A. G., Young, A. W., Michie, J. M., and Whitacre, J., Leaves and stems of turnip greens as a source of some nutrients, *Food Res.,* 11, 432, 1946.

Olson, O. E., Burris, R. H., and Elvehjem, C. A., A preliminary report of the folic acid content of certain foods, *J. Am. Diet. Assoc.,* 23, 200, 1947.

Orgell, W. H., Vaidya, K. A., and Hamilton, E. W., A preliminary survey of some midwestern plants for substances inhibiting human plasma cholinesterase in vitro, *Proc. Iowa Acad. Sci.,* 66, 149, 1959.

Parsons, C. S., Effects of temperature and packaging on the quality of stored cabbage, *Proc. Am. Soc. Hort. Sci.,* 74, 616, 1959.

Pederson, C. S., Mack, G. L., and Athawes, W. L., Vitamin C content of sauerkraut, *Food Res.,* 4, 31, 1939.

Pederson, C. S. and Fisher, P., The Bactericidal Action of Cabbage and Other Vegetable Juices, New York Agricultural Experiment Station, Technical Bulletin 273, Geneva, N. Y., 1944.

Pederson, C. S. and Robinson, W. B., The quality of sauerkraut preserved in tin and glass, *Food Technol.,* 6, 46, 1952.

Pederson, C. S., The influence of temperature on sauerkraut fermentation, *Food Packer,* 37(6), 26, 1956.

Pederson, C. S. and Albury, M. N. The Sauerkraut Fermentation, New York Agricultural Experiment Station Technical Bulletin 824, Geneva, N. Y., 1969.

Peryam, D. R. and Seaton, R. W., Food Consumption and Preference Under Conditions of Restricted and Non-restricted Feeding, Interim Report, Armed Forces Food and Container Institute, Quartermaster Food and Container Institute for the Armed Forces, Quartermaster Corps, U.S. Army, Chicago, 1962.

Petering, H. G., Stubberfield, L., and Delor, R. A., Studies on the guinea pig factor of Wulzen and Van Wagtendonk, *Arch. Biochem. Biophys.,* 18, 487, 1948.

Pooles, C. F., Heinze, P. H., Welch, J. E., and Grimball, P. C., Differences in stability of thiamine, riboflavin, and ascorbic acid in cabbage varieties, *Proc. Am. Soc. Hort. Sci.,* 45, 396, 1944.

Proctor, B. E. and Goldblith, S. A., Radar energy for rapid food cooking and blanching and its effect on vitamin content, *Food Technol.,* 2, 95, 1948.

Pyke, M., The vitamin content of vegetables, *J. Soc. Chem. Ind. (Lond.)*, 61, 149, 1942.

Quinn, V. P., Scoular, F. I., and Johnson, M. L. Ascorbic acid content of cabbage salads, *Food Res.*, 11, 163, 1946.

Ranganna, S. and Setty, L., Nonenzymatic discoloration in dried cabbage. Ascorbic acid-amino acid interactions, *J. Agric. Food Chem.*, 16, 529, 1968.

Reder, R., Ascham, L., and Eheart, M. S., Effect of fertilizer and environment on the ascorbic acid content of turnip greens, *J. Agric. Res.*, 66, 376, 1943a.

Reder, R., Spiers, M., Cochran, H. L., Hollinger, M. E., Farnish, L. R., Gieger, M., Sheets, O. A., Eheart, J. F., and Carolus, R. L., The effects of maturity, nitrogen fertilization, storage, and cooking on the ascorbic acid content of two varieties of turnip greens, Southern Cooperative Series Bulletin, 1, 1, 1943b.

Reeve, R. M., Changes in tissue composition in dehydration of certain fleshy root vegetables, *Food Res.*, 8, 146, 1943.

Salunkhe, D. K., Gamma radiation effects on fruits and vegetables, *Econ. Bot.*, 15, 28, 1961.

Salunkhe, D. K., Dhaliwal, A. S., and Boe, A. A., N^6 Benzyladenine as a senescence inhibitor for selected horticultural crops, *Nature (Lond.)*, 195, 724, 1962.

Salunkhe, D. K., Chiang, J., and Singh, B., Water quality determines quality of canned products, *Utah Sci.*, 32(1), 18, 1971a.

Salunkhe, D. K., Wu, M., Wu, M. T., and Singh, B., Effects of telone and nemagon on essential nutritive components and respiratory rates of carrots (*Daucus carota* L.) roots and sweet corn (*Zea Mays* L) seeds. *J. Am. Soc. Hort. Sci.*, 96(3), 357, 1971b.

Salunkhe, D. K., Wu, M. T., and Singh, B., The nutritive composition of pea and sweet corn seeds as influenced by *S*-triazine compounds, *J. Am. Soc. Hort. Sci.*, 96(4), 489, 1971c.

Salunkhe, D. K., unpublished data 1972-73, 1973.

Schwimmer, S., Alternation of the flavor of processed vegetables by enzyme preparations, *J. Food Sci.*, 28, 460, 1963.

Sedky, A., Stein, J. A., and Weckel, K. G., Efficacy of added ascorbic acid the control of discoloration of kraut, *Food Technol.*, 7, 67, 1953.

Self, R., Casey, J. C., and Swain, T., The low-boiling volatiles of cooked foods, *Chem. Ind. (Lond.)*, p. 863, 1963.

Sheets, O. A., Leonard, O. A., and Gieger, M., Distribution of minerals and vitamins in different parts of leafy vegetables, *Food Res.*, 6, 553, 1941.

Sheets, O. A., McWhirter, L., Anderson, W. S., Gieger, M., Ascham, L., Cochran, H. L., Speirs, M., Reder, R., Edmond, J. B., Lease, E. J., Mitchell, J. H., Fraps, G. S., Whitacre, J., Yarnell, S. H., Ellet, W. B., Moore, R. C., and Zimmerly, H. H., Effect of fertilizer, soil composition, and certain climatological conditions on the calcium and phosphorus content of turnip greens, *J. Agric. Res.* 68, 145, 1944.

Sheets, O. A., Petmenter, L., Wade, M., Gieger, M., Anderson, W. S., Peterson, W. J., Rigney, J. A., Wakeley, J. T., Cochran, F. D., Eheart, J. F., Young, R. W., and Massey, P. H., Jr. The nutritive value of collards, *S. Coop. Ser. Bull. No. 38*, p. 5, 1954.

Sheets, O., Permenter, L., Wade, M., Anderson, W. S., and Geiger, M., The effects of different levels of moisture on the vitamin, mineral, and nitrogen content of turnip greens, *Proc. Am. Soc. Hort. Sci.*, 66, 258, 1955.

Singh, K. D., Morris, E. R., Regan, W. O., and O'Dell, B. L., An unrecognized nutrient for the guinea pig, *J. Nutr.*, 94, 534, 1968.

Singh, B., Vadhwa, O. P., Wu, M. T., and Salunkhe, D. K., Effects of foliar application of *S*-triazine on protein, amino acids, carbohydrates, and mineral compositions of pea and sweet corn seeds, bush bean pods, and spinach leaves, *J. Agric. Food Chem.*, 20(6), 1256, 1972.

Smith, E. B., Hiltz, M. C., and Robinson, A. D., Variability of ascorbic acid within and between cabbage, *Food Res.*, 13, 236, 1948.

Somers, G. F., Kelly, W. C., and Hamner, K. C., Changes in ascorbic acid content of turnip leaf dics as influenced by light, temperature, and carbon dioxide concentration, *Arch. Biochem.*, 18, 59, 1948.

Somers, G. F. and Kelly, W. C., Influences of shading upon changes in the ascorbic acid and carotene content of turnip greens as compared with changes in fresh weight, dry weight, and nitrogen fractions, *J. Nutr.*, 62, 39, 1957.

Speirs, M., Miller, J., Whitacre, J., Brittingham, W. H., and Kapp, L. C., Number of turnip green plants for determination of ascorbic acid content, *S. Coop. Ser. Bull.*, 10, 18, 1951.

Steigmann, F. and Shulman, B., The time of healing of gastric ulcers: implications as to therapy, *Gastroenterology*, 20, 20, 1952.

Stevens, H. P., Preliminary study of conditions affecting the nutritive values of dehydrated vegetables, *J. Am. Diet. Assoc.*, 19, 832, 1943.

Stevens, M. A., personal communication, 1973.

Stewart, J. K. and Barger, W. R., Effects of cooling method, pre-packaging and top-icing on the quality of Brussels sprouts, *Proc. Am. Soc. Hort. Sci.*, 83, 488, 1963.

Strehler, V. E. and Hunziker, K., Behandlung von Magendarmgeschwuren mit Kohlsaft Und Bananenfrapps (antiulcusfaktor, vitamin U), *Schweiz. Med. Wochenschr.*, 84, 198, 1954.

Tarkhov, V. S. and Martishenya, A. I., Qualitative investigation of the principal products of army rations in anti-scorbutic vitamin C, *Voen.-Med. Zh.*, 4, 130, 1933, *C. A.*, 29, 8062, 1935.

Thomas, M. H., Brenner, S., Eaton, A., and Craig, V., Effect of electronic cooking on nutritive value of foods, *J. Am. Diet. Assoc.*, 25, 39, 1949.

Van den Berg, L., Changes in pH of some frozen foods during storage, *Food Technol.,* 15, 434, 1961.

Van Duyne, F. O., Chase, J. T., Owen, R. F., and Fanska, J. K., Effect of certain home practices on riboflavin content of cabbage, peas, snap beans, and spinach, *Food Res.,* 13, 162, 1948.

Van Riter, I. G., Acceptance of twenty-six vegetables, *J. Home Econ.,* 48, 771, 1956.

Van Zante, H. J. and Johnson, S. K., Effect of electronic cookery on thiamine and riboflavin in buffered solutions, *J. Am. Diet. Assoc.,* 56, 133, 1970.

Walker, J. C. and Foster, R. E.,, The inheritance of ascorbic acid content in cabbage, *Am. J. Bot.,* 33, 758, 1946.

Wilcox, E. B. and Neilson, A. B., Effect of quantity preparation on the ascorbic acid content of cabbage salad, *J. Am. Diet. Assoc.,* 23, 223, 1947.

Wilson, A. K., Effect of water quality on physico-chemical and organoleptic characteristics of selected canned fruits and vegetables, Ph.D. Dissertation, Utah State University, Logan, 1971.

Wood, M. A., Collings, A. R., Stodola, V., Burgoin, A. M., and Fenton, F., Effect of large-scale food preparation on vitamin retention: cabbage, *J. Am. Diet. Assoc.,* 22, 677, 1946.

Wu, C. H. and Fenton, F., Effect of sprouting and cooking of soybeans on palatability, lysine, tryptophan, thiamine, and ascorbic acid, *Food Res.,* 18, 640, 1953.

Yearbook Statistical Committee, Agricultural Statistics 1972, U. S. Department of Agriculture, Washington, D. C.

Zepplin, M. and Elvehjem, C. A., Effect of refrigeration on retention of ascorbic acid in vegetables, *Food Res.,* 9, 100, 1944.

Zink, F. W., N^6-benzyladenine, a senescence inhibitor for green vegetables, *J. Agric. Food Chem.,* 9, 304, 1961.

Zscheile, F. P., Beadle, B. W., and Kraybill, H. R., Carotene content of fresh and frozen green vegetables, *Food Res.,* 8, 299, 1943.

DEVELOPMENTS IN TECHNOLOGY AND NUTRITIVE VALUE OF DEHYDRATED FRUITS, VEGETABLES, AND THEIR PRODUCTS

Authors: **D. K. Salunkhe**
            J. Y. Do
            Department of Nutrition and Food Science
            Utah State University
            Logan, Utah

            H. R. Bolin
            Western Regional Research Center
            United States Department of Agriculture
            Berkeley, California

INTRODUCTION

Historical

The drying of fruits and vegetables was practiced long before Biblical times by Chinese, Hindus, Persians, Greeks, and Egyptians. The ancient Hindus and Chinese dried herbs, fruits, and vegetables by the sun and wind five thousand years ago. Dates, figs, apricots, and raisins were sun-dried by those early inhabitants of the Middle and Near East. Potatoes were dried by the Inca Indians two thousand years ago by a natural freeze drying process (Feustel et al., 1964). The product known as "chuno" was made by allowing potatoes to freeze and thaw, and squeezing the juice out of the thawed potatoes with bare feet, and repeating this process until they were sufficiently dry. Another dried product of potatoes made by the Peruvian Indians was called "tunta" and it was sun-dried. Sun-drying is still used for many fruits and vegetables. The household preservation of fruits and vegetables by sun-drying has been common in many cultures and civilizations.

According to Brekke and Nury (1964), approximately one third of the tonnage of dried fruits produced in the United States is now dehydrated mechanically, whereas nearly all were sun-dried a few years ago. Apples, prunes, and sulfured grapes are dehydrated, while unsulfured grapes, apricots, peaches, and pears are still mainly sun-dried, however. From the late 1870's to the early 1890's, large quantities of peaches were mechanically evaporated in Delaware and perhaps in some of the older peach growing areas, according to Beattie and Gould (1917). The evaporation techniques applicable to apples were developed in the early years of the 20th century in New York State (Gould, 1907). He described kiln evaporators that consisted of a two story building with a furnace in the bottom room that provided the heat for drying apple slices spread on the slotted floor of the upper room. A cabinet dryer heated with a steam radiator was in use in New York at that time. How to bleach apple slices in fumes of burning sulfur was also described by Gould (1907). The first record of the artificial drying (dehydration) of vegetables appeared in the 18th century (Prescott and Proctor, 1937). The British seem to have pioneered the hot water blanching. The stability of antiscorbutic activity in dried products was improved by scalding the vegetables prior to drying.

The rapid improvement in machinery and drying methods gave a great impetus to the dried fruit industry and contributed to a substantial increase in production (Langworthy, 1912). For example, apple slices were dehydrated by means of hot air with fans; therefore, the drying time was shortened. Purees of peaches or plums were also made into dried products at that time. The advance of dehydration technology both in the U.S.A. and abroad was further accelerated by World War I. The number of dehydration plants in Germany increased from 3 in 1898 to 1,900 in 1917 (Prescott, 1919). In 1919 there were about 25 dehydration plants in the United States (Prescott, 1919). Pugsley (1917) mentioned 12 drying plants in the Midwest region. The Pacific Coast States emerged as the leading area in dried fruit production after World War I and improved dehydrators were being introduced for operation. The new equipment developed in California for drying prunes and grapes was described by Cruess (1919) and Cruess et al. (1920). The new Oregon tunnel with forced heated air circulation and partial recirculation of the air was described by Weigand (1923).

According to Medriczky (1960), modern industrial potato dehydration originated in 1894 and 1902 in Germany to provide a supply of stable dried potatoes for the alcohol fermentation plants when fresh potatoes were not available. Some 5,500 tons of dehydrated vegetables, including green beans, cabbage, carrots, celery, potatoes, spinach, sweet corn, turnips, other vegetables, and soup mixtures, were processed in the United States as reported by the 1920 U.S. census. A few years later, onion and garlic dehydration was started in California and by 1941 the production was well established. The products have been largely used for remanufacture in a variety of other processed foods.

Nichols and Christie (1930) described the use of stack and kiln evaporators and tunnel dehydrators in drying apples and mentioned that the products were essentially the same when evaporation and dehydration were properly accomplished. Eidt (1938) also outlined kiln and tunnel dehydrators used in Nova Scotia for dehydration of apples. He reported that excellent results were obtained by sulfuring apple rings for 30 min in heavy sulfur fumes, and the advantages of short tunnel with a finishing chamber relative to kiln evaporation.

World War II again stimulated production, research, and development in fruit and vegetable dehydration. At that time the U.S. fruit and vegetable dehydration industry was mainly concentrated in the Pacific Coast States. In 1943, 139 vegetable dehydration plants in the United States produced 115 million lb of dried products which were worth nearly 50 million dollars (Von Loesecke, 1955). By the end of the war, the number of companies that dehydrated vegetables in the United States was greatly increased. The manual on dehydration of fruits and vegetables by the U.S. Department of Agriculture (1944) brought the knowledge of the science and technology of dehydration of fruits and vegetables up to date. Instructions for locating a plant, equipment and building requirements, operating and quality control methods, packaging and storage procedures for handling of the dehydrated fruits and vegetables were described in detail in this valuable manual. A bulletin on the principles and equipment for fruit dehydration was published by Perry et al. (1946).

Dehydrated fruits and vegetables have been an important food supply in the United States as well as in the other parts of the world for military operations in major wars. Unfortunately, dehydrated vegetables were disliked by the soldiers because of inferior quality as compared to the fresh counterparts.

The vegetable dehydration industry recovered swiftly from the cutbacks experienced at the end of the war. The production in 1960 reached 166 million lb of dehydrated potatoes and 76 million lb of other vegetables (Anon., 1962). Among the dehydrated potato products, instant mashed potatoes have made great progress and have become popular among consumers. The advancement of vegetable dehydration was attributed to the new technological findings (Tressler, 1956; Feinberg et al., 1964).

During the past several decades manufacturers of machinery, the food industry, government, and university laboratories have been engaged in research and development of new and improved dehydrated fruits, vegetables, and their products to meet the ever-increasing demands of the space age as well as the general consumers who have become much more nutrition conscious. Several new techniques of dehydration have been successfully established and applied in commercial practice.

Significance

Modern technology has greatly increased the yields of fruits and vegetables. One fourth of the highly perishable produce harvested, however, is never consumed due to spoilage during its storage, transportation, and processing. The annual loss is over a billion dollars in the United States alone.

Proper processing and preservation of harvested produce minimize post harvest losses and thus help offset shortages in supply. Preservation of fruits and vegetables by dehydration is highly effective and practicable. Among all the possible food preservation methods, dehydration especially commended itself to military planners because of its space- and weight-saving possibilities.

Remarkable progress has been made in dehydration technology during the past several decades. Some of the proliferating dehydration techniques have been employed by the industry on large-scale production. Others need further improvement to make them economically feasible. The storage stability of dehydrated fruits and vegetables is affected by many factors: temperature, humidity, light, packaging method, microorganisms, trace elements, etc. The dehydrated product should be protected as much as possible from these adverse conditions so that its quality can be maintained.

It is also important that the dehydrated fruits, vegetables, and their products be nutritious. The arrival of fabricated foods in recent years has rendered the fortification of nutrients and improvement of flavors, colors, and textures relatively easy.

DEVELOPMENTS IN TECHNOLOGY

Pretreatments of Raw Product

To obtain a dried product of excellent quality, the raw produce must be harvested and handled properly. Sanitation, storage stability, and retention of flavor and nutritive qualities require that most foods to be dried undergo some pretreatments, such as washing, trimming, slicing, blanching, dipping, or sulfuring.

Harvesting

To obtain high quality dried fruits and vegetables the raw product must be harvested at optimum maturity and processed as carefully and rapidly as possible. Pears are one fruit that is held after picking because the best product is obtained

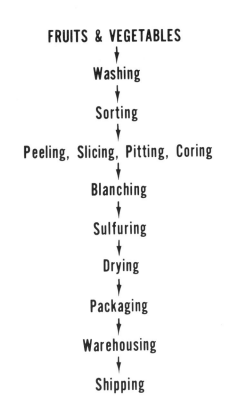

FRUITS & VEGETABLES
↓
Washing
↓
Sorting
↓
Peeling, Slicing, Pitting, Coring
↓
Blanching
↓
Sulfuring
↓
Drying
↓
Packaging
↓
Warehousing
↓
Shipping

FIGURE 1. Flow sheet for dehydration of fruits and vegetables.

when they are picked green and allowed to ripen in storage. Fruits to be dried are usually picked by hand, but ever-increasing labor costs are catalyzing the development of more mechanical harvesting methods. Some bruising is, however, unavoidable with mechanical procedures and this may be accentuated when the fruits are transported in bulk to the processing plant in deep beds. Prunes are now harvested mechanically with the resultant fruit undergoing less physical damage than the hand harvested fruit (Miller, Fridley, and McKillop, 1963).

Vegetables, especially the leafy types, and peas, tend to deteriorate rapidly after harvesting. Rates of respiration and transpiration accelerate in all produce after harvesting and during transportation. Therefore, harvested produce should be delivered to the plant and must be processed as rapidly as possible. Icing in the field or hydrocooling treatment usually delays deterioration.

Washing

Fruits are usually rinsed in cold water at the drying plant to remove surface dust particles and spray residues. Vegetables should be thoroughly washed with particular care taken to remove adhering soil from the root crops by use of heavy sprays and rotary washers.

Peeling and Slicing

Root vegetables, apples, and sometimes clingstone peaches are peeled prior to drying. This is accomplished by various means: abrasion, refractory, lye, hot brine peelers, high pressure steam for root crops, mechanical knife peelers for apples, and dilute lye for clingstone peaches. Recently, a "dry" caustic process came into wide acceptance for potatoes, and is also being applied to other products (Graham et al., 1969). This peeling process has the advantage of using a minimal amount of water, thus reducing pollution problems.

Root crops then are cut into cubes, julienne strips, or slices for drying; cabbage is shredded; and potatoes may be diced and dried into powder form. Prunes, grapes, berries, and cherries are dried whole, although the latter may be pitted. Apricots, nectarines, and peaches are halved and pitted; pears are halved; and apples are peeled, cored, and either sliced or sectioned. Early in the season unpeeled apples are sometimes sliced and dried for use in products like apple butter.

Dipping

Alkaline Dip

Dipping, which involves immersion of the product in an alkaline solution prior to drying, is used primarily for fruits that are dried whole, especially prunes and bleached grapes. The process cannot be used for berries, with the possible exception of cranberries. Dipping facilitates drying by forming fine cracks in the skin. A sodium carbonate or lye solution (0.5% or less) is used at a temperature usually ranging from 93.3 to 100°C. Concentration, temperature, type of dipping solution, and duration of immersion time vary with the condition of the fruit being treated.

In Australia and some Mediterranean countries cold dip solutions such as carbonate or lye with olive oil or commercial dipping solutions are used. The main active ingredients of commercial solutions are oleate esters (Stafford et al.,1973). These dips accelerate moisture loss by causing the wax platelets on the grape skin to dissociate, thus facilitating water diffusion (Ponting and McBean,

1970; Possingham, 1972). Raisins produced by cold dip processes are light in color.

Acid Dip

Mrak et al. (1942) used an acid predip before sulfuring to provide a product of better color stability. A 1% ascorbic acid and 0.25% malic acid dip have been used to retard enzymatic browning in peaches (Anon., 1972). Other researchers have used an acid dip instead of sulfuring as a method for obtaining a light colored product (Lazar et al., 1963). However, this product would undoubtedly have to be held at reduced temperatures to keep it from darkening during storage.

Sulfuring or Sulfiting

Fruits

Sulfur dioxide treatments are widely used in the food industry (Roberts and McWeeny, 1972). Cut fruits and grapes (for producing golden-bleached raisins) are exposed to sulfur dioxide gas before drying. Apples may be dipped in a sulfurous acid solution prepared by dissolving sodium bisulfite or sulfur dioxide in water. The sulfur dioxide absorbed by the fruit is intended to maintain an attractive color, prevent spoilage, and preserve certain nutritive attributes until marketed. Sufficient sulfur dioxide must be absorbed by the prepared material to allow for losses that occur during drying and subsequent storage.

Fruits are sulfured by placing them in an adjustable vented compartment containing burning sulfur. The sulfuring houses can be of different construction, such as PVC tents, wood frame cells, or cement block tunnels (McBean et al.,1967). In addition to this initial sulfuring all cut fruits are sulfured a second time immediately after processing, just before packaging. This resulfuring is usually accomplished with burning sulfur. However, a faster and easier method used recently has been to dip the processed fruit for 30 sec in a 5 to 7% potassium metabisulfite solution (Stafford and Bolin, 1972).

The absorption and retention of sulfur dioxide are influenced by the temperature, duration of sulfuring, and concentration of sulfur dioxide, as well as the size, condition, maturity, and cultivar product being sulfured. Immature fruit is characterized by an increased absorption but less retention than fully mature fruit. Higher sulfuring temperatures tend to reduce absorption, but greatly facilitate the retention of sulfur dioxide. Losses are greater in sun-drying than in conventional dehydration. The levels of sulfur dioxide advisable in dried fruits that are to be used domestically depend on the duration of storage and the storage temperatures. Some optimum levels in parts per million are apricots, 3,000; peaches and nectarines, 2,500; pears, 2,000; apples, 1,500; and golden-bleached raisins, 1,000. The maximum amount of sulfur dioxide that dried fruits for export can contain varies with the importing country. Some countries allow entry of any dried fruit as long as it does not contain over 2,000 ppm sulfur dioxide; others have specific restrictions for each fruit.

Vegetables

Treatment of vegetables with sulfur dioxide gas is impractical. Sulfite solutions are preferred as the most practical method of controlling absorption. The absorbed sulfur dioxide is held much more tenaciously by the more neutral vegetable than by the acid fruit, and the retention in the dried product is therefore more directly related to the concentration of the sulfite.

Cabbage, potatoes, and carrots are routinely sulfited prior to dehydration. The highest levels are required for cabbage, for which limits of 750 to 1,500 ppm have been set. Limits of 200 to 500 ppm are imposed for potatoes and carrots. The sulfite treatment unquestionably eases production problems by lessening the danger of scorching as the product approaches dryness. The value of sulfiting for carrots has been questioned and another pretreatment, starch dipping (or spraying), holds promise for preparation of large quantities.

For small-scale operations blanched vegetables may be dipped either in continuous or batch operation for a specified time in a solution with concentration controlled and maintained. Spraying, however, is the most practical method, and its application preferably accomplished in the blancher.

Blanching

Blanching consists of a partial cooking, usually in steam or hot water, prior to dehydration. Foods have been experimentally blanched, however, using hot air and also infrared and microwave radiation (Ralls et al., 1973; Huxsoll, Dietrich, and Morgan, 1970). The heating denatures enzymes in the product. The degree of enzyme inactivation indicates the effectiveness of the blanching treatment. The activity of polyphenol oxidase is followed in fruits, that of catalase in cabbage, and that of peroxidase in other vegetables.

Commercially, continuous blanchers are favored over the batch type, which involves a 2 to 10 min exposure to live steam. Series blanching in hot water is advocated in some countries, with the solids content of the water maintained at an equilibrium level to minimize leaching losses. Blanching can produce one or more of the following effects:

a. reduces drying time

b. removes intercellular air from the tissues

c. causes softer texture

d. retards the development of objectionable odors and flavors during storage by enzyme inactivation

e. retains carotene and ascorbic acid during storage

f. removes pungency, particularly in onions or

g. causes loss of soluble solids.

The expulsion of air from the tissues has two effects. In vegetables, if the tissues are collapsed, the cells beneath the surface are protected effectively from the adverse effects of oxygen in the air. This is particularly noticeable in products high in starch, such as potatoes. Possibly this is one reason why potatoes are considered a good vegetable to dry. In fruits such as apricots, peaches, and pears, it imparts the desired translucent appearance to the dehydrated product.

Blanching, as well as dehydration, can cause an increased crystallinity of the cellulose in the product. This crystallinity change evidently causes a change in the texture of the fruit or vegetable (Holdsworth, 1971).

Individual Quick Blanching (IQB)

A modification to steam blanching was described by Lazar et al. (1971) as the Individual Quick Blanch (IQB) process, a method that produces less effluent than conventional steam blanching. A heating step and a holding step are comprised in the IQB process. Heating is done in a condensing steam unit with raw products one layer deep, and heat is applied to raise the temperature of the product to inactivate enzymes (above 87.7°C). In the adiabatic holding step after the pieces leave the heating section, the pieces achieve a uniform temperature and are allowed enough time to inactivate enzymes and to yield the desired texture. Lazar et al. (1971) also reported that "preconditioning" carrots prior to blanching by warming and partial drying could significantly reduce effluent. Partial drying provided a surface that absorbed condensing steam, and prewarming reduced the sensible heat required for blanching. Bomben et al. (1973) compared the IQB with and without preconditioning to conventional steam blanching in green beans, lima beans, Brussels sprouts, and green peas using a pilot plant IQB system consisting of a warming and drying preconditioning unit, a steam chamber for single layer belt heating, a deep moving-bed holding chamber for temperature equilibration and enzyme inactivation, and an air-water spray cooling unit. It was found that vegetables blanched with IQB with heat preconditioning showed no discernible difference in quality as those blanched under commercial conditions, and there also was a substantial reduction in the product solids loss in the effluent.

Dehydration Procedures
Sun-drying

Sun-drying is the most ancient method of drying foods and is still in use in many parts of the world. It is by far the cheapest source of heat for the removal of water from fruits and vegetables; however, with sun-drying goes the risk of losses due to inclement weather and the difficulty of maintaining a high degree of sanitation. Various fruits, vegetables, and their products can be sun-dried. Apricots, nectarines, peaches, pears, and currants are some fruits that are sun-dried. The sound, ripe fruit, after being pretreated as described in the previous section, are loaded on trays and placed in the sun-drying yard, remaining there until the fruit is about two thirds dry. Then the trays are stacked to allow the later stages of drying to proceed slowly in the shade. Most of the grapes for producing raisins are sun-dried on paper trays placed directly in the rows of the vineyard.

Fruit Leathers

Sun-drying, or forced air dehydration, can also be adopted to prepare fruit leathers. In this process, a puree is first made from the respective fruit — apricots, peaches, etc.; if fresh fruit is used, a heating step is advisable to inactivate the enzymes. The pureed product can be lightly sweetened if desired and is finally spread in a thin layer on a plastic film and dried. The dried product has a bright translucent appearance, chewy texture, and a good fruit flavor. These leathers can be stored by rolling them up while they are still on the film, covering with more plastic, and sealing. The fruit leathers retain their color and flavor for at least a month at room temperature, four months under refrigeration, and a year when frozen (Anon., 1972a). If bisulfite is incorporated into the formulation, their storage life can be greatly increased.

Excellent information on details of sun-drying equipment and field layout including location, sanitation, harvesting, transportation, pretreatment, drying, and storage of various fruits is available from Mrak and Phaff (1949) and Cruess (1958).

Cross-flow Drying
Tunnel Drying

Despite the increasing popularity of continuous dryers of various types, tunnel dryers are still widely used by the food industry in the United States because their simplicity and great versatility are unmatched by any other type of dryer. Foods in pieces of almost any size and shape can be successfully dried.

Tunnel dryers, as used for dehydrating fruits and vegetables, have been extensively studied and thoroughly described in many technical publications. Particularly important are those of Perry et al. (1946), Van Arsdel (1951a, b), Kilpatrick et al. (1955), and the U.S. Department of Agriculture (1959).

A tunnel dryer is basically a group of truck-and-tray dryers, operated in a programmed series so as to be semicontinuous. Truckloads of freshly prepared material are moved at intervals into one end of the long, closely fitting enclosure, the string of trucks is periodically advanced a step, and the dried truckloads are removed at the other end of the tunnel. The hot drying air is supplied to the tunnel in any of several ways: counter-flow, concurrent or parallel-flow, center exhaust, multistage, and compartment arrangements. Gentry, Miller, and Claypool (1965) determined that parallel-flow drying was a faster method of drying. Drying by this method can increase the capacity of a dehydrator tunnel by 37% over the counter-flow procedure; consequently the fuel consumption will jump about 12%. Occasionally, a combination of drying methods is employed. Apricots are sometimes dried by first putting the sulfured cut fruit in the sun for a few hours to develop color and then putting them in a tunnel dehydrator for more rapid final drying (Miller, 1965). This procedure permits a faster cycling of trays during the drying season, thus allowing a grower to dry a greater tonnage using a given number of trays in a given drying yard area. Another drying method is the dry-blanch-dry (DBD) process, which was developed by Lazar et al.(1961, 1963) for dehydration of apricots, peaches, pears, grapes, and apples to produce a translucent, quick cooking, and excellently flavored dried product. In this process the fruit is first washed, halved, and pitted. It is then sulfured lightly by a dip in bisulfite solution and dried to 50% weight reduction in a conventional tunnel dehydrator. After blanching in a steam blancher, the fruit is returned to the tunnel to finish dehydrating. The apricots processed this way resemble sun-dried apricots in appearance and have a pleasing, fruity flavor.

Cabinet Drying

Cabinet dryers are arranged for batch operation and are usually held at a constant temperature, though humidity may decrease during the process of drying. Air flow may be across or through the trays with or without recirculation. Cabinet dryers are flexible as to type and size of operation, although production may be lower and labor requirements greater than for a tunnel dehydrator. Beavens (1944a, b, and c) has described cabinet

DBD

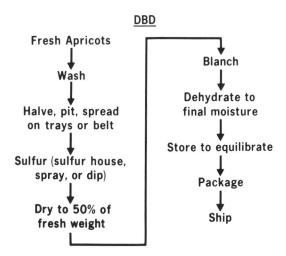

FIGURE 2. Flow sheet for dry-blanch-dry. (Source: Lazar et al., 1963.)

dryers suitable for small scale operations. Fresh air enters the cabinet, is drawn by the fan through the heater coils, filtered by screens to remove dust, and then blown across the materials on trays to an exhaust system. The air may pass across and between the trays or may be directed up through the perforated trays.

Continuous Conveyor Drying

The continuous conveyor dryer consists of an endless belt that carries the material to be dried through a tunnel of warm circulating air. This drying method has the advantage of essentially automatic operation, which minimizes labor requirements.

These conveyor dryers are being used extensively in food dehydration plants in the United States, and have been recently applied to materials that have conventionally been dried on trays in tunnels, such as apple, beet, carrot, onion, potato, and sweet potato pieces. The primary reason for the changeover has been the saving in costs of loading and unloading the dryer and of maintaining the drying surface in good condition. Final quality is also better in some products. The conveyor dryer is best adapted to the large-scale drying of a single commodity for the whole operating season; it is not well suited to operations in which the raw material or the drying conditions are changed frequently, because extensive hours of startup and shutdown make it difficult to produce satisfactory products (U.S.D.A., 1959).

Pneumatic Drying

The method of drying finely divided solids while they are being conveyed in a hot air stream has been used for the first stage or rough drying of granulated cooked potatoes in the manufacture of potato granules. The partial drying desired takes place in a few seconds for finely granulated wet potato, whose initial moisture content generally has been reduced to around 35 to 40% by an admixture of the fresh mash with recycled dry product ("add-back"). Most of the material is finer than 60 mesh, but a small proportion may be 20 mesh or coarser. The moist granulated material is usually sucked into the conveying pipeline at a low pressure throat ("venturi"). Air velocity in the pipeline is just enough to overcome the free fall velocity of the coarser particles to keep them suspended, while avoiding unnecessary abrasion damage to the delicate potato cells. The pipe is sufficiently long to give the necessary retention time at that air velocity. Ordinarily, the conveying pipe is vertical, rather than horizontal, so that the product may not tend to settle down and drift along the bottom of the pipe.

The conveying section is generally terminated by a diverging section which helps to separate the light, completely dry particles from the heavier, still somewhat moist ones, and gives the latter more time in the hot air. The conveying riser ends with a cyclone collector where the granular solid is separated from the outgoing airstream.

Through-flow Drying
Bin Drying

Through-flow drying is widely used as a finishing step in the dehydration of piece-form vegetables when they constitute a deep bed of granular, nonsticky material, just as it has been used for a long time to reduce the moisture content of wet grain to a safe storage condition.

The usage of bin dryers in certain vegetable dehydration plants is described in the Management Handbook (U.S.D.A., 1959). Bin dryers are used, particularly for piece-form vegetable products, to complete the drying operation after most of the

moisture has been removed in a tunnel dryer or the equivalent. Typically, they reduce the moisture content of a partially dried, 10 to 15% moisture cut vegetable to 3 to 6% or even lower in the case of onion slices and possibly cabbage shreds.

Broad features of the design and operation of bins, as operated in the finishing stage of vegetable dehydration, were described in the Dehydration Manual (U.S.D.A., 1944). A characteristic unit consists of a metal or wooden box equipped with an air inlet at the bottom and a wire mesh deck or false bottom with an air supply duct below it, so arranged that warm dry air can be passed through the nearly dry product piled on the deck.

Belt Trough Drying

This is a type of continuous through-flow drying that can be used for a wide range of fruits and vegetables that have a firm texture in piece form. The belt trough dryer was put into commercial operation in 1957. It consists of a wire-mesh belt running on sprockets mounted on three drive shafts. During part of its path of travel, the belt forms a trough, whose bottom rests on an inclined, flat-surfaced air grate. The material to be

FIGURE 3.   Belt trough dryer.

dried is continuously cascaded in this trough while a current of hot air blows up through the grate and bed of material. High operating temperatures result in high drying rates without causing appreciable heat damage to vegetable pieces because the constant agitation keeps any individual piece from being exposed to the very hot dry air for more than a moment. Then each piece is surrounded by air at much lower temperatures for a longer time before again moving into the zone of intense drying (Lowe et al., 1955; U.S.D.A., 1959; Van Arsdel and Copley, 1963).

Fluidized Bed Drying

In this type of drying, heated air is blown up through the food particles with just enough force to suspend the particles in a gentle boiling motion. Semi-dry particles such as potato granules gradually migrate through the apparatus until they are discharged dry. Heated air is introduced through a porous plate that supports the bed of granules. The moist air is exhausted at the top. This process is continuous and the amount of time that the particles remain in the dryer can be regulated by the depth of the bed and other means. This type of drying can be used to dehydrate grains, peas, and other particulates.

Fluidized bed drying has numerous advantages, such as:

1. High drying intensity.
2. Uniform and closely controllable temperature throughout.
3. High thermal efficiencies.
4. Time duration of the material in the dryer may be chosen arbitrarily.
5. Elapsed drying time is usually less than in other types of dryers.
6. Equipment operation and maintenance are relatively simple.
7. The process can be automated without difficulty.

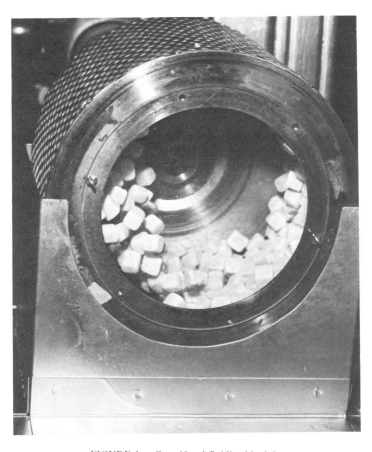

FIGURE 4.   Centrifugal fluidized bed dryer.

8. Fluidized dryers are compact and relatively small.

9. Several processes may be combined in a fluidized bed dryer.

A centrifugal fluidized bed dehydrator has been developed which allows the use of greater air velocities (Farkas et al., 1969; Lazar and Farkas, 1971; Brown et al., 1972). Fluidization of carrot dice and other particles in a centrifugal force field up to 10 times that of gravity greatly extended the practical mass flow range. Uniform dense beds were produced without slugging problems. Heat transfer at moderate gas temperature can be improved by increasing the gas velocity, and large pressure drops across the grid supporting the bed are not needed to obtain smooth fluidization.

Carrots, potatoes, and green beans in piece forms have been successfully blanched and dehydrated by this method. Particulate foods with homogeneous starchy cell structure such as potato, carrot, squash, and sweet potato in diced form are easily puffed by the centrifugal fluidized bed method.

Freeze drying has also been investigated in relation to fluidized bed drying. Freeze drying is constantly being explored and it produces a good quality product. High costs, however, have limited its application. Atmospheric fluidized bed freeze drying is one possible solution. Malecki et al. (1970) determined that this type of drying was impractical for apple juice because of the slow drying rate (0.6% moisture loss/hr) of the frozen juice droplets at their caking point temperature of -34°C. However, atmospheric fluidized bed freeze drying should be feasible where diffusion distances are very short as in foods of open structure (Malecki et al., 1970).

Atmospheric freeze drying has been used by Lewin et al. (1962) to effectively dry piece form foods when they dried 8 mm carrot dice to 15% of its original weight (8.5% moisture) in 24 hr at -20°C. Woodward (1963) reported drying times and temperatures for 90% water removal from several meats and vegetables freeze dried at atmospheric pressure. O'Meara (1963) discussed problems associated with atmospheric pressure freeze drying.

Drum Drying

Drum dryers are designed to handle solutions, fruit and vegetable purees, pastes, and sludge materials. Drum drying processes are used successfully in the production of powdered cranberries, tomato cocktail, pea and bean soups, apple flakes, potato flour, tomato juice flakes, mashed potato flakes, and sweet potato flakes.

A drum dryer consists of one or two hollow drums which are fitted so that a heating medium, usually steam but occasionally water or a special, high temperature, heat-transfer liquid, can be circulated through them. The drums are mounted to rotate about the symmetrical axis and are customarily driven with a variable speed mechanism. A feeding device applies a thin, uniform layer of the material to be dried on the hot drum surface. A knife or doctor blade is fitted to the drum at an appropriate location. The material is dried as the heated drum rotates toward the doctor blade which scrapes the thin layer of dry material from the drum surface.

Cording et al. (1957) studied the relations among solids content of a potato mash, drum speed, steam pressure, product moisture content, rate of product output, and bulk density of dry sheet leaving the drum. A mash containing 80% moisture can be reduced to 5% moisture in about 20 sec. The air surrounding the drum, containing much water vapor, is usually moved into an exhaust stack by natural ventilation, but sometimes is exhausted mechanically so that the humidity at this point can be controlled at a desired level. Even though products like tomato powder have been produced for over 30 years with the aid of drum dryers, they are still being improved. Lazar and Miers (1971) added a low humidity sample collecting shroud to the dryer, thus increasing its efficiency and producing a better flavored powder.

Fruit Cloth, Bars, and Molds

Bolin, Fuller, and Powers (1973) developed a versatile product called apricot "cloth," which was produced by drying apricot concentrate on a double drum dryer operating 3 to 4 rpm and 130°C, with a clearance of 0.4 mm. The concentrate was pretreated with 0.5% sodium bisulfite to protect the carotene content and the light color. The drum-dried material contained about 12% moisture and had a bright orange color and a good apricot flavor. This cloth can be cut into small strips, packaged, amd marketed as is. Multilayers of the cloth can be pressed and molded into

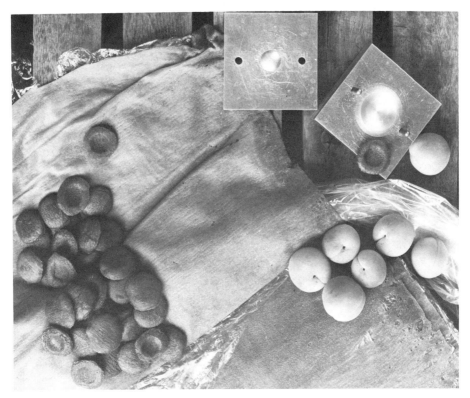

FIGURE 5. Sheets of "apricot cloth," the mold used to shape the dried apricot, molded apricot halves, and fresh apricots. (Source: Bolin et al., 1973.)

apricot bars or apricot halves, which have a shape and taste similar to sun-dried apricots.

One advantage in the molded apricots is the formulation possibilities in which protein powders, vitamins, or other nutrients can be added either to the apricot concentrate before drying or sprayed on the sheets of cloth before lamination and pressing.

The atmospheric drum dryer is most common in the food industries, although the vacuum drum dryer has been used for certain specialty products.

Spray Drying

Spray drying has been widely used because drying is almost instantaneous. The liquid product to be dried is atomized into a hot air steam and the particles as they dry fall into a collector. The design, operation, and use of spray dryers in food industries are discussed in detail in a review by Seltzer and Settlemeyer (1949).

Spray drying has many attractive features, such as rapid drying, continuous production, large through-put, simplicity of operation, and low labor cost. The main disadvantages of the process

have been the likelihood of heat damage to the material and the occasional production of too high a proportion of very fine powder.

Puff Drying
Explosive Puff Drying

Explosive puff drying, which has long been used for preparing rice and wheat breakfast cereals, has been recently expanded to include fruit and vegetable pieces. For a product to be explosive puff dried it must be partially dried to a moderately low moisture level. The material to be puffed is then sealed in a heavy metal cylinder which is pressurized by the application of heat. The pressure is instantaneously released by opening one end of the cylinder. This sudden change in pressure causes the food to expand greatly. This treatment yields the product a more porous structure, allowing for faster final drying and more rapid rehydration. Eisenhardt et al. (1962) used this method to prepare dehydrated potato and carrot pieces. The cooking time of the dried materials was shortened from the usual 15 to 30 min to about 5 min. After such a short time

FIGURE 6. Batch type vacuum shelf dryer.

cooking in boiling water, the diced pieces assumed their original shape and were tender and palatable.

According to a U.S.D.A. pilot test, introduction of carbon dioxide into the puffer retarded non-enzymatic browning of potatoes in explosion-puffing during dehydration and the optimum amount is 0.35 lb/min of gas to 0.65 lb of steam (Anon., 1973).

Vacuum Puff Drying

"Vacuum puff drying" refers to the dehydration of liquid materials in vacuum by evaporating water under pressures in excess of 1 mm of Hg so that the products puff during drying and form expanded, honeycombed structures. This process is distinguished from freeze drying in that the latter is done under extremely low pressures, always well below 1 mm and sometimes as low as 5 μm, with the materials in a frozen state.

In puff drying of juices, a porous structure is obtained by applying vacuum to a viscous juice concentrate. Bubbles of water vapor and entrapped air form and expand throughout the material. Vacuum, temperature, and viscosity of the concentrate are controlled so that the puffed material has about 20 times the volume of the liquid concentrate. In certain juices such as tomato, which have low soluble solids content and a high insoluble solids or pulp content, puffing is not satisfactory unless air or another gas is incorporated into the concentrate.

One of the early references to puff drying is in the patent granted to Heyman (1943). He described a method of producing a solid, dehydrated citrus product that involved mixing citrus fruit materials with corn syrup, expanding the mix under heat and vacuum, and dehydrating the mix to form an expanded dehydrated cellular product having intercommunicating cells. Hayes and co-workers (1946) reported on the puff drying of 50° brix orange juice concentrate in a laboratory vacuum oven at the extremely low pressure of 100 μm. These powders reconstituted readily and were judged superior to canned juices but not as good as fresh juice.

Schroeder and Cotton (1948), Schwarz (1948), and Schwarz and Penn (1948) reported on further studies in which the liquid was sprayed on a plate. Within a short time the film would swell and puff as much as 3 in. The vacuum would range from 600 μm down to 300 μm or less at the end of the drying cycle. Drying from a liquid film doubled the production rate compared with a frozen film. Drying temperatures under 48.9°C were recommended.

Sluder et al. (1947) described a continuous belt dehydrator designed for orange juice drying. An endless stainless steel belt was housed within an evacuated chamber. The concentrated orange juice in the form of thin film puffed while drying by radiant heat and conduction in the evacuated chamber. The powder was vacuum packed without exposure to air. Strashun and Talburt (1953) reported adapting the puff drying technique to continuous dehydration of 100% orange and tomato juice powders.

Freeze Drying

Freeze dehydration (lyophilization) consists of freezing the product and then evaporating (sublimating) the water, which is in the form of ice, directly into the vapor phase with no liquid interim. Water vapor is removed from the frozen food, which is achieved by reducing the pressure in a sealed drying chamber by a vacuum pump to the extent of 50 to 1,000 μm of mercury. This pressure holds the fruits and vegetables below their eutectic temperature, facilitates rapid evaporation of water molecules from the frozen state, and condenses the vapor either within the chamber or in a separate chamber. Heat energy applied to the frozen product causes the water molecules to

break loose and migrate to the block of ice on the condenser, where their energy is removed by cooling. Heat may be applied to the frozen product by conduction or radiation, or both.

Freeze dehydration produces one of the highest quality food products obtainable by any drying method. The porous, nonshrunken structure of the freeze-dried product facilitates rapid and nearly complete rehydration when water is added. The low processing temperatures and the rapid transition minimize the extent of various degradative reactions such as nonenzymatic browning, protein denaturation, and enzymatic reactions that are associated with other drying methods, and also help reduce the loss of volatile flavor and aroma. Freeze drying as related to mass transfer and process developments has been reviewed by Karel (1973).

Accelerated Freeze Drying

Accelerated freeze drying (AFD), a process developed by the British Ministry of Agriculture, Fisheries, and Food at Aberdeen, uses conductive contact heating. As reported by Hanson (1961) and by Coston and Smith (1962), heat is supplied through the dry layer from expanded metal inserts, which provide contact at some points for heat conduction and, at the same time, an open path for easy vapor escape. The expanded metal inserts fit between the food and the heater plates.

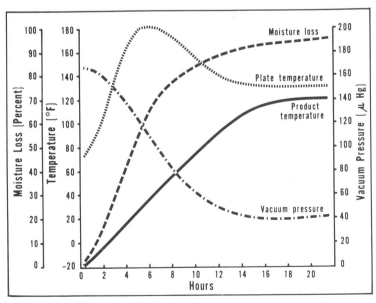

FIGURE 7. Typical drying curves for applesauce show plate and product temperature, vacuum change, and moisture loss in relation to time. (Source: Salunkhe et al., 1971.)

A pressure of approximately 4 psi on the heating plates promotes good thermal contact with the food. A similar process was described by Oldencamp and Small (1965), who proposed supplying heat by conduction from inflatable platens made of a flexible material such as rubber. A heating fluid, such as ethylene glycol, is supplied to the interior of the platens and to heating plates. Again, expanded metal inserts could be used to provide a path for vapor flow. Another scheme for supplying heat conductivity comes from Smithies and co-workers (1959), and involves heated spikes that penetrate the material undergoing freeze drying.

Liquid Nitrogen and Cryogenic Freezing

Fruits and vegetables are generally frozen by a cold air blast prior to freeze dehydration. Cryogenic advances in food technology have brought new techniques utilizing liquid carbon dioxide, Freon, and liquid nitrogen. Liquid nitrogen freezing is done by immersing the respective fruits (raspberries, strawberries, peaches,

and cherries) in the liquefied gas in a stainless steel insulated container (Wolford et al., 1967, 1971). Cryogenic freezing has less adverse effect on texture than air blast freezing because the rapid freezing produces very small crystals (Lee et al., 1966, 1967, 1968).

Liquid nitrogen is at a temperature of about −195.6°C. Fruits and vegetables to be frozen in liquid nitrogen are subjected to this extremely cold temperature for a matter of seconds at which time a hard crust is frozen around the outside of the fruits. The crust in turn freezes the inside of the food item after it is removed from the nitrogen. In many cases, cold gas is used to precool the fruit so that it may be partially frozen prior to actual immersion in the liquid nitrogen. The entire freezing operation is quick, and must be closely controlled.

The smallness of the ice crystals formed, and other factors associated with ultrarapid cryogenic freezing, may account for the superior texture and drip characteristics of various small fruits and fruit segments frozen in liquid nitrogen or dichloro-

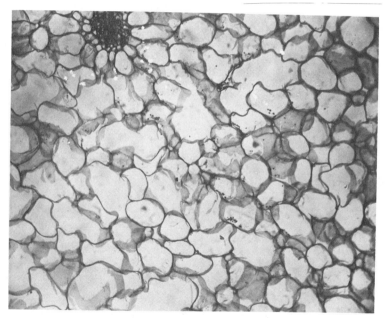

FIGURE 8A.

FIGURE 8.   Photomicrographs of parenchyma cells of apple: a. Fresh, b. Conventional hot-air dehydrated, c. Freeze dehydrated, d. Freeze dehydrated and rehydrated, e. Cryogenic frozen and freeze dehydrated, and f. Cryogenic frozen, freeze dehydrated, and rehydrated. The figures indicate textural changes and integrity of cells during the process of dehydration. Cells of conventional dehydrated products were collapsed whereas freeze dehydrated and cryogenic freeze dehydrated and subsequently rehydrated apple slices maintained their cell integrity and texture rather similar to the fresh sample. (Source: Lee, Salunkhe, Bolin, 1967.)

FIGURE 8B.

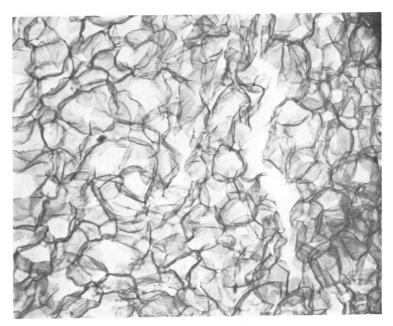

FIGURE 8C.

difluoromethane (Webster and Benson, 1966; Wolford et al., 1967). Recently developed commercial methods for the quick freezing of food, including fruits, involve treating the material with a spray of liquid nitrogen. Figs were frozen in liquid nitrogen and stored for 10 months (Gelashvili, 1971). Best results were obtained at −18°C storage, with the chemical and organoleptic

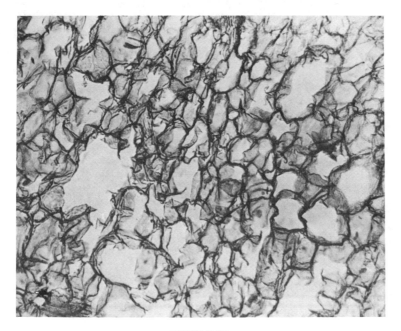

FIGURE 8D.

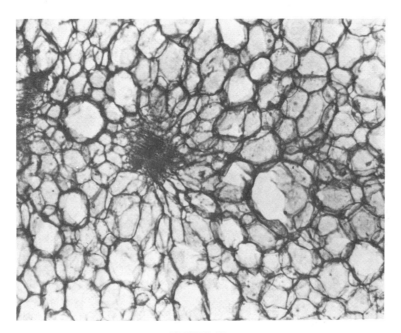

FIGURE 8E.

FIGURE 8F.

properties of the stored defrozen figs being practically the same as of fresh figs. Figs frozen in liquid nitrogen were reported to retain their natural color for a relatively long period.

To minimize the structural damage of fragile pieces of food caused by the cascade of liquid nitrogen during conventional liquid nitrogen freezing, McLain and Abott (1971) devised a system that introduced droplets of liquid nitrogen in the vapor phase instead of or in addition to using the liquid state.

Freeze Concentration

Freezing can also be used to remove water from fruit and vegetable juices to produce a concentrated product. In the process of freezing the single strength juice, all of the components do not freeze at once but some of the water forms ice crystals in the mixture first. The partially frozen slush is centrifuged through a fine mesh screen to collect the concentrated unfrozen solution while frozen ice crystals retained in the centrifuge are discarded. By repeating this process several times on the concentrated unfrozen juice it is possible to increase the final concentration severalfold. Concentrating by freezing exposes the product to mild conditions, thus minimizing nutritional and quality changes. A good discussion on freeze concentration has been written by Thijssen (1970). Freeze concentration has been applied commercially to orange juice. However, the drawbacks for this process are high processing cost and incomplete recovery of concentrated juice occluded to the ice crystals which ends up in losses.

Foam-mat Drying

In this process a juice is whipped to a stable foam after the addition of a small amount of foaming agent and a stabilizer. Subsequently, the foam is spread on a porous tray and dried. No vacuum is used and the cost of such an operation is accordingly very small. The material dries rapidly due to the porous nature of the foam, and since evaporation is fast, high heat can be used without impairing flavor and color. Flavor changes that occur during drying have been studied extensively by Bolin and Salunkhe (1971).

Apple, cherry, and peach juices were concentrated to a powder using a foam-mat process by Morgan et al. (1961) and Salunkhe et al. (1963). Appropriate amounts of solubilized soya protein and methyl cellulose were added to each juice sample, which was then whipped for 5 min and spread in a 1 mm thick layer on perforated stainless steel trays. The trays were placed in a 72°C dehydrator for 25 min, after which they were put in a 15% humidity room to cool. The powder was then scraped off the trays and packaged.

FIGURE 9. Foam-mat dryer.

Microwave Drying

Microwaves are the portion of electromagnetic spectrum between far infrared and the conventional radio frequency region. They are similar to light waves and can pass through, be reflected by, or be absorbed into materials, depending on the material. This type of energy is generated by specially designed radio frequency power tubes from direct current at high voltage. As the energy passes through material such as foods, the molecules within the food attempt to align themselves with the electric field direction. As they oscillate around their axis, heat is produced by the intermolecular friction, with the amount of heat varying depending on structure, shape, composition, and mass of the product.

Two types of power sources are now available for the production of microwave energy for use in food processing. One is a continuous system that operates at 915 Mc using a single 25 kW power source. The other operates at 2,450 Mc, with a series of 2.5 kW power sources.

Another important factor in microwave heating is the depth of penetration. In general, the penetration increases for microwave frequencies as the frequency decreases, but the penetration also depends upon the materials being exposed. The penetration will be infinite in perfectly transparent substances and zero in reflecting materials such as metals. For most absorptive materials such as food, the penetration is finite. Under ideal conditions, nearly all the microwave energy impinging on food is absorbed at frequencies between 1,000 and 3,000 Mc, but below 400 and above 3,000 Mc absorption of the incident energy is substantially reduced. Changes in the composition of food while it is being heated also modify the behavior of the microwave energy. In the material being dehydrated, heating due to incident microwaves causes increasing temperature accompanied by increasing resistivity of the material. As resistivity increases, the rate of absorption of the microwave energy increases and the temperature of the material rises even more rapidly. Crystallization of sugars or salts caused by drying may also produce a conduction ordinarily associated with metals and other low resistance materials (Copson, 1962).

Penetration is quantitized by half-power depth, the thickness of material that reduces the microwave energy to one half the incident level. For water the half-power depth is about one half in. at

56

2,450 Mc and 3 in. at 100 Mc. For many food processes, the 2,450 frequency does not provide adequate penetration. A nonhomogenous substance may exhibit a range of temperatures in a microwave oven and it is important, therefore, that each product be evaluated individually (Nury and Salunkhe, 1968).

The product being processed by microwave heating is in effect its own source of heat. While the product is being heated, the oven chamber itself is cold. The lack of heating within the environment indicates that the energy dispensed is almost totally absorbed by the product being processed. The dimensions of microwave processing chambers are greater than the wavelength used. As a result, the energy, after it is introduced into the chamber, bounces from wall to wall, passing through the product from all directions. Variations in product loading patterns, and in its thickness and water content, influence the processing rate. Microwave energy can be used to process products of different sizes or shapes, subject to depth restrictions. In contrast with other heating methods, microwave energy penetrates significant distances into food materials and thus can speed drying, cooking, baking, etc. (Nury and Salunkhe, 1968).

The use of microwave energy has received only scant attention as an aid for food dehydration. Other forms of heating depend on first transferring heat to the surface of the product, with subsequent slow conduction of heat into the product. In microwave heating the heat input is highly efficient and in a carefully designed tunnel, power utilization efficiencies greater than 70% can be attained (Allaire, 1965; Pollack and Foin, 1960).

Rushton et al. (1945) experimented with 13-Mc radio frequency (RF) heating of vegetable blocks as a way to reduce the respective moisture contents of cabbage and potato from 9 and 15% to 5 and 7%. In air-finish-drying by RF energy the time required was about one fifth of that in a cross-air blow dryer.

Work done by Jeppson (1964) and Nury and Salunkhe (1968) on the finish-drying of fruits and vegetables demonstrated a synergistic effect between hot air and microwaves. Huxsoll and Morgan (1966) reported on the use of microwaves in "puffing" foods. The RF energy has been used for processing potato chips. A continuous tunnel system using microwave energy for the finish frying of potato chips is nearing completion. By frying to about 8% moisture content and then finishing off the dehydration in a microwave oven

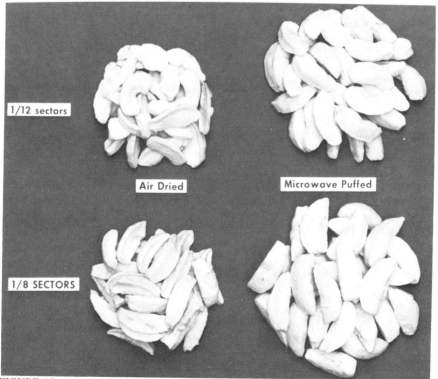

FIGURE 10.   Apples dried by conventional and microwave puff process. Note the puffing effect which may help in rehydration. (Source: Huxsoll and Morgan, 1966.)

at 100°C (instead of the 160°C temperature of the fryer), it is possible to use most varieties of potatoes, regardless of the temperature of storage, without causing excessive browning (Davis et al., 1965a, b, c; Decareau, 1965). Porter et al. (1973) studied the effect of microwave finish-drying of potato chips on their texture, color, and oil content. Two primary variables, the intermediate moisture content of the chips before microwave drying and the reducing sugar content of the raw potatoes, were pointed out. They found that potato chips removed from the oil at intermediate moisture content above 13% were unacceptably tough after microwave-drying. Potatoes with more than 0.9% reducing sugar had to be removed from the oil (160°C) at intermediate moisture above 13% in order to obtain acceptable color after microwave drying. The microwave finish drying raised the limiting reducing sugar content from about 0.4 to about 0.9%, and yielded chips that contained 10% less oil than that of conventional controls. Blau et al. (1965) presented economic and product justification for the use of microwaves at 2,450 Mc in the manufacture of potato chips using the Litton multiple-module system. O'Meara (1966) described the economic savings obtained by using 915 Mc microwaves for the finish frying of potato chips.

A French food machinery company has made a microwave dryer that operates under vacuum. The combinination of microwave heat and vacuum permits short drying times at relatively low temperatures. The result is a powder that reportedly retains the flavor, aroma, and color of the original product. And it is readily soluble, even in cold water (Trauberman, 1973).

Dehydrofreezing and Dehydrocanning

Dehydrofreezing reduces the weight of a fruit or vegetable piece to about one half its original weight by warm air drying before freezing. A wet surface exposed to a hot air current will lose moisture rapidly, resulting in an evaporative cooling effect which will maintain the surface temperatures considerably lower than the temperature of the surrounding atmosphere. Product temperature is thereby maintained below the level that favors deteriorative changes in foods. Drying may be carried out in cabinet, tunnel, continuous conveyor, or continuous mixing dryer.

Dehydrofrozen apples are high quality double strength stocks suitable for use in pies and its commercial production exceeds that of any other dehydrofrozen commodity. The apples are cored, trimmed, sliced, sulfured by immersion, dried to 50% weight reduction, and frozen. Lazar et al. (1961) developed an alternate procedure using little or no sulfur treatment and blanching the product after it is partially dried. Dehydrated apples should be frozen rapidly, preferably airblast frozen, at -29 to -34°C and stored at -18°C or lower.

Dehydrofrozen peas have been a successful market item, primarily for institutional use. The peas are harvested, cleaned, graded, blanched, and slit. Seventy percent of the initial water in the pea is removed by drying and the weight and bulk are reduced by half. Commercially this is accomplished by combining a belt-trough drier and a specially designed concentric double drum drier. As the product leaves the drier, it drops onto a freezing belt where it is quickly frozen. The partially dried product freezes much faster than the undried pea. The resulting power savings in freezing almost offsets the fuel consumption for drying, and substantial economies can be realized in the subsequent costs of packaging, transporting, and storing (Van Arsdel and Copley, 1964). Boggs and Talburt (1952) found dehydrofrozen peas could be equal in quality to frozen or strictly fresh pod peas. Pimentos (peeled) and bell peppers that are to be dehydrofrozen are first cored, trimmed, washed, and diced. The dice are then partially dried. Belt-trough dryers have proved extremely efficient, bringing the moisture content of the pepper dice from approximately 92% to about 77% moisture in 25 min. The dehydrated dice are then preserved by freezing (Van Arsdel and Copley, 1964). By using the dehydrofreezing technique, packing, storage, and shipping costs are reduced two thirds as compared to freshly prepared pepper dice (Rasmussen et al., 1957). Potatoes and carrots are also dehydrofrozen commercially. Dehydrofrozen products of snap beans, lima beans, corn, celery, onions, and sour cherries are experimented with in the laboratory (Lazar, 1968). LaBelle and Moyer (1966) reported that dehydrofrozen cherries have excellent color stability even after holding for several weeks unfrozen at 2°C, and also indicated that pies made from the reconstituted dehydrofrozen cherries were considered acceptable.

Dehydrocanning is similar to dehydrofreezing in producing a canned product. Dehydrocanned apples are a modification of canned apples in which the prepared slices are dried to half their original weight prior to packaging.

Osmosis

Sugar and Syrup Treatment

Osmotic drying consists of removing a percentage of the moisture from a fruit or vegetable piece by placing it in contact with granular sugar or a concentrated sugar solution. The product is reduced to about 50% of its original weight by the osmotic dehydration, after which it may be frozen or dried further in an air or vacuum dryer. The osmotic agents used are various dry sugars, sugar-starch mixtures, and sugar syrups. Sucrose was found to be the best all around dry substance. Ponting et al. (1966a, b) and Lee et al. (1966, 1967, and 1968) described a method of partial drying of fruits by osmosis in sugar or syrup, followed by vacuum drying. The products had a porous crisp texture and retained a large percentage of the flavor volatiles of the fresh fruit.

About 40% of the water can be removed from certain tropical fruits by the osmotic process. For bananas and plantains, this involves immersing slices in a concentrated sugar solution for about 18 hr; the ripe mango requires the same treatment for about 4 hr; by contrast, the green mango must be immersed in a condensed salt solution for about 24 hr. The economics of the process probably depend upon the availability of cheap sugar and on the possibility of using spent sugar solutions in canning, bottling, or soft drink plants. Subsequent drying in the sun or in hot air currents is suggested. A sulfur dioxide treatment is sometimes used to help preserve color. A detailed report of experiments carried out in Ottawa, together with recipes that can be used in processing plants, is available (Hope and Vitale, 1972).

Glycerol Treatment for Celery

Shipman et al. (1972) equilibrated transversely sliced celery sections in glycerol solutions prior to dehydration, and reported that the glycerol treatment, which is similar to osmotic dehydration, produced an improved quality rehydrated product. Further studies by Do and Salunkhe (1973) showed that pretreatment of crosscut celery slices by propylene glycol, ethanol, and glycerol (at 50%) prior to freeze dehydration resulted in better rehydrated celery slices.

Reverse Osmosis

Bolin (1970) prepared reverse osmosis concentrates of apple, cherry, and peach juices by using an experimental unit equipped with two sheets of

OSMOTIC DEHYDRATION

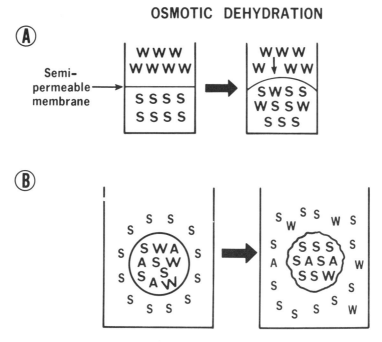

FIGURE 11.   Diagramatic representation of osmotic dehydration. W: water, S: sugar, A: acid.

laboratory prepared cellulose acetate membranes. The juice was circulated and its pressure increased, using an air activated pump, to approximately 1,000 to 1,200 psi.

Fruit juice concentration by reverse osmosis was faster than by the regular osmosis. The water removal rate for cherry juice at 1,000 psi (68 atm) gradually declined over 3 hr from about 90 cc/min to about 10 cc/min. The drop in rate probably occurred because primarily the juice originally contained about 84% water, which would give it a calculated osmotic pressure of about 11 atm. As the water was removed the percent sugar present increased and thus the osmotic pressure.

The actual osmotic pressure of fruit juice, as determined by Morgan et al. (1965), was 15.6 atm for 12° brix and 102 atm for 45° brix juice. The osmotic pressure of a juice increases as it becomes more concentrated. The rate of permeant flow, and hence of concentration, depends on the differential between the applied pressure and the osmotic pressure of the material being concentrated. The greater this difference, the faster the concentration rate. Flow rate could also be influenced by polarization, which was caused by the build-up of concentrate on the membrane, or by clogging of the membrane pores by pectinaceous materials.

Diffusion Membrane

Fruit juices have also been concentrated by a "diffusion membrane" procedure (Bolin and Salunkhe, 1971), where the water in the juice diffuses through a semipermeable membrane and is then evaporated by warm air.

Hot Oil Immersion

Dehydration of fruits and vegetables by hot oil immersion is comparatively new. This drying procedure has numerous advantages over conventional hot air dehydration, one being that the products dry faster. The faster drying is due to a higher temperature during drying, and to the high thermal conductivity of oil (4.5×10^{-4} cal/sec cm C° compared to 5.4×10^{-5} cal/sec cm C° for air). Some products also undergo an improvement in texture, flavor, storage stability, and rehydration characteristics. In addition oil-dried foods have enhanced caloric values because of fat incorporation, which could be beneficial in developing countries. Many developing countries preserve

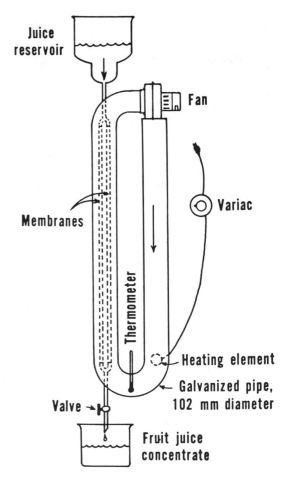

FIGURE 12. Small scale diffusion membrane concentrator. (Source: Bolin and Salunkhe, 1971.)

fruits and vegetables by hot oil immersion because the method is simple and cheap.

Sathiraswasti and Salunkhe (1958) developed a process of deep fat frying various maturities of peas in coconut oil at 177°C. They found that deep fat fried overmature peas (size 6) had better (more "eye appealing") color, higher protein content, and less fat and moisture than did comparably treated and small peas. Since overmature peas are of little value from canning, their utilization in deep fat frying would give processors as well as growers additional income.

Deep Fat Fried Snack Foods

Many deep fat fried snacks can be produced from such vegetables as sweet potatoes, beets, carrots, peas, lima beans (Kelley and Baum, 1955), fruit slices (Salunkhe, 1956), soybeans (Collins and Ruch, 1969), turnips (Reddy et al., 1971), corn on the cob (Collins et al., 1971), and toasted cottonseed kernels (Lawhorn et al., 1970). Also, hot oil immersion is used in India to prepare dehydrated convenience foods (Bhatia, 1970).

A popular deep fat fried snack is the onion ring. Recently, a fabricated "onion ring" was developed wherein chopped onions were mixed with a binding agent, and were extruded, battered, breaded, and deep fat fried (Anon., 1972a). Several advantages over the conventional methods of preparing onion rings were claimed by developers of this approach. It was reported that about 25% of the onion rings sold at retail as of January, 1972, were made by their process. Collins and Sanders (1973) indicated that chopped soybeans may be combined with chopped onions and binder, coated with batter and bread, and prepared as a tasty deep fat fried snack food with improved protein content.

One of the most popular deep fat fried snack foods in the U.S. is potato chips. Potato chips are a high energy food produced by the rapid dehydration of potato slices in direct contact with hot fat at temperatures ranging from 162.8 to 190.6°C. During dehydration, which requires about 4 min, enough fat is absorbed by or on the chip to constitute 30 to 50% content of the finished product. This absorbed fat adds to the flavor and nutritive value of the chip. The chip moisture content is approximately 2% as it emerges from the fryer. Finally, 1 to 2.75% salt is added to complete a ready-to-eat food.

Freeze Leaching and Quality Improvement

A new concept for minimizing oil absorption into a product has been introduced by Nonaka, Weaver, and Hautala (1972). This process consists of incorporating a surface freezing of potato strips and subsequently leaching sugars in hot water prior to frying. By this treatment in their work with french fries, oil absorption was reduced about 25% and the product had better than average texture and color.

Sweet Potato Chips

A high quality sweet potato chip that retained the characteristic yellow-orange color and had crisp texture and good flavor was produced by Hoover and Miller (1973) by optimizing a combination of treatments prior to hot oil immersion. These treatments included blanching the raw strips in a water solution containing 0.5 to 0.75% sodium acid pyrophosphate for 2 min, followed by

partial dehydration in a continuous forced-air dryer at 104.4°C and then deep fat frying. The optimum oil temperature was from 183.3 to 154.4°C and the required heating time varied with the moisture in the blanched strips. The moisture content of the finished sweet potato chips was 3 to 4%.

Azeotropic Drying

This new method utilizes the fact that water forms low boiling point azeotropic mixtures with numerous organic solvents. The food product proceeds continuously into a tank being fed with ethyl acetate; the mixture is then moved to a reactor with circulation to which a vacuum is applied. The vacuum removes the azeotrope, which boils at 24°C under a vacuum of 100 mm Hg and –19°C at 25 mm Hg. Enough residual solvent is removed in a vacuum oven at 38°C to meet the U.S. Food and Drug Administration's requirements regarding ethyl acetate. The process occurs at temperatures that preclude water flashing and cellular destruction, and the final product is of comparable, and often better, overall quality than is freeze-dried material. Since the solvent penetration accelerates the water diffusion rate, it can be used for pieces of material large enough to require much longer periods of time for freeze drying (e.g., whole fruits and vegetables). The flavors and colors of some products are then removed by the solvent, and it is necessary that these are reintroduced at a later stage.

Various solvents can be used, such as ethyl alcohol. The ethyl alcohol, however, requires a larger ratio of solvent/material, i.e., 30 cc/g, compared to 15 cc/g for ethyl acetate. The process can be applied to a frozen product with equally satisfactory results. Similarly, ethanol and ethanol/water mixtures at –30°C have been used. The technique involves immersing the material in the solvent at –30°C, and after the water has interacted with the solvent the product is submitted to a blast of dry air (Anon., 1968; Nakamura, 1967).

STORAGE STABILITY

Browning Reactions

During processing as well as during storage, deleterious changes can occur in fruits and vegetables which affect their appearance, flavor, and odor. The major changes that occur result in the enzymatic or nonenzymatic development of the brown pigmented materials.

Enzymatic Browning

Enzymatic browning takes place fast, following a mechanical injury sustained during handling and processing of fresh fruit and vegetable tissues, because they contain active enzymes. Enzymes such as polyphenoloxidase (also called polyphenolase or phenolase) in the ruptured cells get mixed with oxygen and the substrate and catalyze the oxidation of colorless phenolic compounds into o-quinones (red to reddish-brown) which then combine with amino acid derivatives to form highly colored complexes. These enzymes, which include catecholase, tyrosinase, ascorbinase, and polyphenoloxidase, are common in fruits and vegetables. Therefore, as these foods are prepared for subsequent dehydration, undesirable browning can occur unless the cut products are treated in some manner. The main treatments consist of inactivating the enzymes with heat or sulfur dioxide.

Nonenzymatic Browning

Nonenzymatic browning occurs in dried foods during protracted storage. Numerous theories have been advanced for the mechanism of this reaction. Stadtman (1948) reviewed some of these, including the Maillard condensation theory.

Hodge (1953) investigated reactions in a model system containing typical components of foods. He noted three kinds of nonenzymatic browning:

1. Carbonyl amino acid-type which includes the reaction of aldehydes, ketones, and reducing sugars with amines, amino acids, peptides, and proteins.
2. Caramelization-type which occurs when polyhydroxy carbonyl compounds (sugars and polyhydroxycarboxylic acids) are heated in absence of amino compounds.
3. Oxidative reaction-type in which ascorbic acid and polyphenols are converted into di- or polycarbonyl compounds. These reactions may or may not be enzyme-catalyzed, nonenzymatic browning.

Song and Chichester (1967) reported on the kinetic behavior and mechanism of the Maillard Reaction where chemical reactions between the sugars and amino acids present in the fruits and vegetables are involved. The initial reaction takes place between the aldehydic group of a sugar and the amino group of a protein or amino acid, followed by a complex reaction scheme which leads to melanoidin formation. Sulfur treatment can prevent the initial condensation reaction by forming nonreactive hydroxy sulfonate sugar derivatives.

Schiff base ⇌ N-substituted glycosylamine ⇌ (Amadori rearrangement) N-substituted 1-amino-1-deoxy- → 2-Ketose

→ Reductones ⇌ Dehydroreductones → Melanoidins

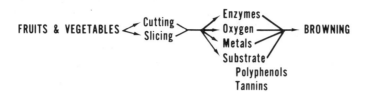

FIGURE 13. Simplified schematic presentation of enzymatic browning of cut fruits and vegetables.

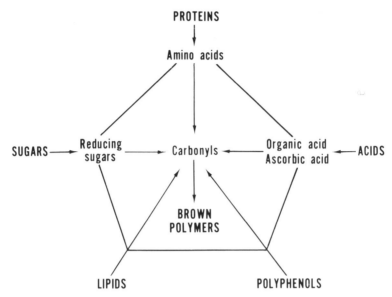

FIGURE 14. Simplified schematic presentation of nonenzymatic browning reactions occurring during dehydration processes of fruits and vegetables.

In caramelization, heating of sugars produces hydroxymethyl furfural, which polymerizes easily. This reaction may be slowed by bisulfite, which reacts with sugars to decrease the concentration of the aldehydic form.

Similar nonenzymatic browning may arise as the ascorbic acid in fruits and vegetables decomposes to the easily polymerizable furfural. The effectiveness of sulfur treatment in preventing this browning may be attributed to the reactivity of bisulfite toward carbonyl groups existing in the breakdown products.

In summary, the brown-pigmented products associated with nonenzymatic browning are not produced by a single reaction but are the results of a multiplicity of reactions.

Factors Affecting Storage Stability
Sulfur Dioxide

Cut dried fruits and some vegetables are treated with sulfur dioxide so they will maintain their light, natural color during extended storage. Within limits, darkening rates, especially for dried fruits, are inversely proportional to the products' sulfur dioxide content (this generalization applies to products that have been sulfured moderately). Therefore, any conditions that accelerate sulfur dioxide loss will in turn accelerate the darkening of the product. This has been illustrated by Schrader and Thompson (1947) and Bolin and Boyle (1972).

Different predip treatments have been studied as to their effectiveness in increasing sulfur dioxide retention. Apricots dipped in a 5% sodium citrate solution before sulfuring retained more sulfur dioxide during drying than did untreated controls. Alkaline dips using phosphate or bicarbonate buffers have also effectively increased sulfur dioxide retention. Potatoes treated with calcium ion in conjunction with sulfite had more protection from browning than did those treated with sulfite alone (Simon et al., 1955).

Temperature

Dried fruits, vegetables, and their products should be stored at relatively low temperatures to maximize their storage life. During storage, produce not only turns brown but also gives off carbon dioxide and absorbs oxygen. The transfer rate of these gases out of and into the product quadruples for every $10°C$ rise in temperature. Therefore, storage temperature is of vital importance in relation to the maintenance of quality in dried foods.

Light

Light is detrimental to dehydrated fruits and vegetables. Bolin et al. (1964) studied the effect of prolonged illumination on dehydrated apricots, peaches, and apples in transparent films stored at $32.2°C$. They found that the sulfur dioxide loss and the rate of darkening in peaches and apricots were almost unaffected by light; however, the content of β-carotene, which is the light-sensitive precursor of vitamin A, was reduced. However, light did increase the rate and amount of sulfur dioxide loss and a product darkening in apples.

Packaging Materials and Package Environment

The type of packaging to be used varies with expected storage conditions. Packages can be divided into two major groups, rigid and flexible, with each group having distinct advantages and limitations.

Rigid containers include metal cans and glass jars. These offer physical protection and insure against gas or vapor transfer. Freeze-dried foods are currently packaged this way by filling the containers with the product, evacuating, back-flushing with nitrogen to less than 2% oxygen (to reduce oxidative rancidity, to maintain high concentration of ascorbic acid, and to inhibit oxidative browning), and finally sealing. Gasket material on the lids differs from that required for cans to be retorted, since freeze-dried products are not further heat processed.

Small sized cans can be vacuum packed. Most large cans, however, will collapse at the vacuum required for product stability (27 or more inches of mercury). Nitrogen packaging can be done in two ways. In one method, a hole is drilled in the can lid and the can is closed in the usual way. The can is next placed in a vacuum chamber, evacuated, and back-flushed with nitrogen, and then tip soldered. More commonly the filled cans are run through a sealing machine that applies only the first clinch. The partially sealed cans are placed in a vacuum chamber, evacuated, flushed with nitrogen, and sealed completely. Vacuum and gas packaging of cut fruits extend their storage life, especially the lower moisture products.

Substitution of nitrogen headspace for air had little effect on the rate of anthocyanin degradation in freeze-dried strawberry puree, but did reduce the degree of browning (Erlandson and Wrolstad, 1972). Wong et al. (1956) reported that in-package desiccation used with an inert atmosphere appears to be the most favorable packaging procedure with regard to the stability of a vacuum-dried tomato juice powder stored at various temperatures under a variety of packaging conditions. Sapers et al. (1973) reported the flavor of nitrogen-packed potato flakes was not affected by storage for 12 months at $23°C$. In a study of chemical changes in freeze-dried mushrooms, Luh and Eidels (1969) reported that for better quality retention the products were recommended to be packed in aluminum-film-combination (AFC) pouches under nitrogen, and stored at lower temperatures, preferably below $20°C$.

Flexible packages maximize the use of available packaging space, and also they are usually cheaper than rigid types. Products in pouches can be either packed atmospherically or by vacuum. If ·the proper films are used, the packages are impervious to moisture, light, and oxygen. Films used for flexible packaging where low vapor transfer is desired are customarily 3-ply laminates, such as a polyolefin-foil-mylar.

Moisture

The moisture content of a product has a decided influence on its storage stability. Hendel et al. (1955) in their work on dried potatoes indicated that their storage stability was increased when their moisture content was reduced from 7 to 4%. Dehydrated sweet potatoes maintained their quality longer when their moisture level was maintained below 4% (Molaison et al., 1962). There are critical water activities for some products below which browning is minimized. Rockland (1969) found that some foods have an optimum moisture level, above or below which their storage life is diminished.

Water Activity

Water activity (a_w) is defined as the ratio of the vapor pressure of water (P) in the food to the vapor pressure of pure water (P_o) at the same temperature, $a_w = P/P_o$. High moisture foods are soft and flaccid and must be heated or chemically treated to inhibit microbial growth. The moisture content is over 30% and water activity (a_w) is .9 and up. Intermediate moisture foods are semi-moist, firm, and more susceptible to Maillard reaction, and less susceptible to fat oxidation than low moisture dry foods. The water activity (a_w) is .85 and moisture content is about 20 to 30%. Low moisture (dried) foods are hard and firm. They are resistant to microbial deterioration and less susceptible to Maillard reaction; they are more susceptible to fat oxidation than intermediate moisture foods. The water activity (a_w) is .7 and below, and moisture content is below 20% (Brockmann, 1973).

In-package Desiccant

Fruits and vegetables can either be completely dried to a low moisture level mechanically or the largest percentage of the water can be removed and then the product packaged with a material that will continue to absorb moisture during storage. In-package desiccation is achieved with packets of calcium oxide or silicas which are placed in a hermetically sealed container with the product. For example, orange juice crystals placed in containers at 3% moisture level can be brought down to less than 1% moisture content. Use of in-package desiccants to bring moisture content to 1% or lower should permit storage of dehydrated fruits and vegetables for 6 months at 21.1 to 37.8°C without significant loss of vitamins. Fumed silicas can be used to prevent agglomerating or caking of powdered foods, dried convenience foods, spray-dried tomato powder, and orange juice crystals. When used at a rate of 1 to 2% in powdered foods, silicas will prevent moisture pickup after the seal of the package has been broken, thereby eliminating caking. These chemically inert, hydrophilic powders can absorb up to 40% of their weight without losing free flow characteristics (Anon., 1973a).

Silicas may be used: (1) in foods in which they have been demonstrated to have an anticaking effect, at levels not to exceed what is reasonably required to produce the intended effect, (2) at rates not to exceed 2% by weight of the food, and (3) in baby foods only as an anticaking agent in salt and salt substitutes that are components of these foods.

Antimicrobial Treatment

One of the main reasons for drying food is to lower its water content to a level that cannot support the growth of microorganisms. In general, bacteria will not grow if the water activity is below 0.91, yeasts will not grow if it is below 0.88, and molds 0.80. However, proliferation of these organisms is also affected by other factors, such as the type of food, microbial load, treatment during processing, etc.

Dried fruits have recently been packed at high moisture levels so that a tender, moist, palatable, ready to eat product could be produced. To insure against microbial spoilage at these higher moisture levels, the products require an additional treatment. Nury et al. (1960) evaluated a number of preservatives for use on dried fruits finding that a potassium sorbate dip was the most effective treatment of the 35% moisture prunes tested. This treatment also provided a residual effect in that the prunes did not support microbial growth even in opened packages. As the moisture level of a product increases, the effectiveness of a particular antimicrobial agent decreases (Bolin and Boyle, 1967). In some products, like dates, where it is harder to inhibit microbial growth, advantage can be taken of the synergistic effect provided by using a combination of treatments, such as potassium sorbate-methyl bromide (Bolin et al., 1972).

Trace Elements

Some salts and metals are detrimental to the nutritive value, flavor, and storage quality of dehydrated fruits and vegetables. The raw materials for dehydration may be exposed to them during washing and pretreatment steps. Calcium has a firming effect on the texture. Sulfur treatment (sulfiting) may prevent the browning of dehydrated products. Iron and copper compounds combine with tannins to cause blackening and may accelerate the degradation of ascorbic acid. Sodium, magnesium, and calcium sulfates impart a bitter flavor. Certain salts of zinc, cadmium, and chromium have toxic effects (Salunkhe et al., 1971; Chiang et al., 1971).

ation is a highly acceptable process for ...p.....vation and reduction of weight of fruits, vegetables, and their products. Compression of dehydrated foods to decrease their volume and thus reduce packaging, handling, storage, and transportation costs has become increasingly important, especially for military usage. Various dehydrated fruits and vegetables have been compressed and subsequently reconstituted to their normal appearance and texture through rehydration.

Compressed dehydrated cabbage and carrots in block forms were produced in the United Kingdom during World War II (Gooding and Rolfe, 1957). Most freeze-dehydrated fruits and vegetables can be compressed with little or no fragmentation when properly preconditioned by spraying the dehydrated product with plasticizers such as water, glycerol, or propylene glycol before compression.

Rahman et al. (1969, 1970, 1971) reported that freeze-dehydrated peas, corn, sliced onions, spinach, carrots, green beans, blueberries, and red tart pitted cherries were successfully compressed with compression ratios ranging from 1:4 to 1:6. Several hundreds to 1,000 lb/in.^{2} of pressure has been used to compress the various freeze-dehydrated products; therefore, some textural destruction is bound to occur which then affects the rehydration property. Freeze-dehydrated, compressed products of fruits and vegetables that will rehydrate rapidly to the original product are needed.

NUTRITIVE VALUE

Dehydrated fruits, vegetables, and their products are good sources of energy, minerals, and vitamins. Some provide moderate amounts of protein to the diet. Recent advances in technology have opened a new era in developing various new fabricated products by utilizing dehydrated fruits and vegetables in manufacturing items like powdered drinks, extruded snacks, compressed bars, etc. These well-designed fabricated foods that have been fortified with vitamins and vegetable proteins are going to be able to provide the daily nutritional demands of our bodies.

Dehydrated foods are at least equally as nutritious as other processed foods. In the study of Thomas and Calloway (1961), cabbage, carrots, corn, and green beans were dehydrated both in the raw and cooked states by exposure to heated air. Cabbage contained a large amount of ascorbic acid, and this was well retained in the dehydrated products. As prepared for serving, the β-carotene in all products studied was highest in the dehydrated samples. No difference between dehydrated and normal foods was found with respect to digestibility or support of nitrogen balance and body weight in man, nor were they noted to alter gastrointestinal irritability.

DeGroot (1963) compared the nutritive quality of the proteins in 12 different cooked food items with that of the corresponding cooked products after dehydration by incorporating the item in diets as the sole source of protein in an animal feeding experiment, and determined the true digestibility and biological value of the protein. It was reported that after hot air dehydration the biological value of cabbage, lima beans, and green beans, and the digestibility of red beans were somewhat lower than in the nondehydrated counterparts, whereas hot-air dried kale and turnip greens and freeze-dried corn showed no noticeable change in protein quality. The general conclusion was that hot air dehydration tends to cause less protein damage as compared to spray drying or freeze drying.

Nutritional Losses

Dehydrated fruits and vegetables are concentrated in nutrients. However, during the process of dehydration some vitamin potency may be lost. The extent of vitamin destruction will depend on the care exercised in preparing the material before dehydration, on the dehydration process used, and on the conditions of storage of the dehydrated product. Vitamins A and C are sensitive to oxidative degradation, thiamine can be destroyed by heat and sulfuring, and riboflavin is light sensitive.

Pretreatment

Fruits as well as some vegetables contain significant quantities of vitamin C and carotene (provitamin A) and smaller amounts of B vitamins. Some loss of vitamin content is inevitable during drying. Sulfur dioxide is a strong reducing agent therefore, it protects oxygen sensitive vitamins, such as A and C, and minimizes their loss during processing. The presence of sulfur dioxide enhances the retention of vitamin C to the extent

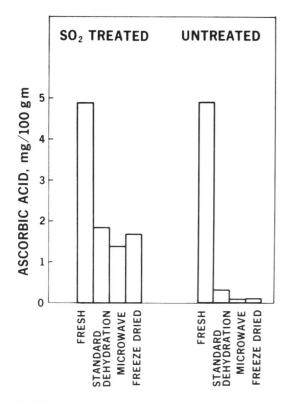

FIGURE 15. Ascorbic acid content of treated and untreated dehydrated Winesap apples. Ascorbic acid is retained better if apples were treated with SO$_2$ prior to dehydration. (Source: Nury and Salunkhe, 1968.)

that 50% of the original content may be retained in the dried product (Morgan and Field, 1929, 1930). On the other hand, thiamine is totally destroyed by SO$_2$ (Morgan et al., 1935). Losses of carotene during drying may be as low as 10 to 15%.

Blanching

Blanching is an essential step to minimize nutrient loss, especially in vegetables. As much as 80% decrease in the carotene content of vegetables will occur if they are processed without enzyme inactivation. The best commercial methods where the product is adequately blanched will result in only about 5% carotene loss, depending upon the particular product.

Stevens (1943) found that ascorbic acid losses during blanching varied with the product being processed; sweet potatoes, rutabagas, and carrots only lose 10 to 20%, while peas and white potatoes lose about 60%. Also, white potatoes did not lose any thiamine during blanching, but peas and white corn lost about 50%.

Ascorbic acid is easily lost during blanching because it is water-soluble and heat sensitive. Ralls et al. (1973) found that spinach blanched by the hot gas procedure retained 34 mg ascorbic acid/100 g compared to 21 mg for the hot water blanched product.

Blanching and freezing steps that are used in conjunction with some types of drying can affect the vitamin content of the final product. Schroeder (1971) indicated that vegetables lost 40% of their vitamin B$_6$ and 47% of their pantothenic acid, and fruits lost 15% and 7%, respectively, from the blanching-freezing operation.

Drying

Kohman (1942) stated that vitamin C suffers almost complete destruction in most dehydrated products, and that dehydrated vegetables retain their vitamin values poorly. Tressler (1942) made one of the first reviews of the literature dealing with the nutritive value of dried and dehydrated fruits and vegetables. More recently, Labuza (1973) reviewed the effects of dehydration and storage on the protein and vitamin contents of foods, indicating that the major loss of fat-soluble vitamins, such as vitamins A and E, is probably due to the reaction of peroxides and free radicals, which are oxidation products of lipids, with these vitamins. Therefore, time, temperature, and oxygen availability are critical with respect to vitamin loss.

Stevens (1943) determined that onions and peas did not lose any ascorbic acid during drying, but that corn and sweet potatoes lost over 60%. Thiamine was not affected by drying; however,

FIGURE 16. Simplified schematic presentation of oxidation of fats and oils during dehydration.

carotene was, with yellow corn and green beans losing 25% and carrots and sweet potatoes over 60%. Moyer (1943) found that potatoes that contained 6.2 mg thiamine/100 gm only contained 4.8 mg after blanching and drying.

Fruits can be sun-dried, dehydrated, or processed by a combination of these methods. Sun-drying causes large losses in carotene and ascorbic acid. Salem and Hegazi (1973) determined that in sun-drying of apricot juice, a large part of the carotenoids and ascorbic acid was lost during the processing and drying. In addition, the quantities of certain amino acids such as lysine, arginine, and threonine were diminished. However, since some fiber was removed, some of these materials could have been lost in this way.

Vegetables artificially dehydrated or sun-dried tend to lose their nutrients in the same order of magnitude as fruits. Thiamine content reduction is about 15% in blanched tissues, while unblanched may lose 75%. Rapid drying retains greater amounts of ascorbic acid than does slow drying. Generally, the vitamin C content of vegetable tissue will be lost during a slow sun-drying process. Dehydration, especially spray drying and freeze drying, reduces these losses.

TABLE 1

Neutral and Acidic Free Amino Acids in Sulfured Winesap Apples Dehydrated by the Three Methods (Basic amino acids were not found in measurable quantities)

Amino acids	Standard dehydration	Microwave-dehydrated	Freeze-dried
	$(\mu m/g$ of apples x $10^{-3})$		
Aspartic acid	70.7	16.3	86.8
Threonine	5.77	1.05	4.10
Serine	15.6	2.63	19.8
Glutamic acid	9.35	3.15	40.15
Proline	5.77	–	4.16
Glycine	3.64	1.05	0.88
Alanine	54.1	18.4	21.3
Valine	6.24	0.89	6.20
Methionine	0.88	(trace)	0.42
Isoleucine	5.77	1.61	6.24
Leucine	3.12	0.89	1.04
Tyrosine	19.2	21.5	1.04
Phenylalanine	4.16	3.68	8.84
Asparagine[1]	99.4	2.63	50.9
Hydroxyproline[2]	8.0	29.04	(trace)
β-Alanine	(trace)	2.1	2.1

[1] Asparagine peak may have contained glutamine as well.
[2] Traces of lysine were also found.
(Source: Nury and Salunkhe, 1968.)

Nury and Salunkhe (1968), in their work on microwave drying of apples, found that microwave energy had an influence on the neutral and acidic free amino acids of apples. Apples that had been hot air-microwave-dried had an overall lower free amino acid content than those prepared by straight dehydration or freeze drying.

Subjection of a food item to excessive heat will greatly affect the proteins in the product. Proteins are heat sensitive; therefore, overheating can reduce the protein efficiency ratio (PER) of the particular food, thus making it unavailable to the body. Also, some amino acids, such as lysine and methionine, have highly reactive functional groups which can react rapidly at higher temperatures. Freeze drying would therefore have the minimal effect on protein.

Niacin and riboflavin, two water-soluble vitamins, are fairly stable to heat and oxidation, and they also are not affected by sulfite. Consequently, neither of the vitamins or vitamin B_{12} is lost to any extent during normal hot air dehydration.

Storage

Nutrient loss during storage is largely dependent on storage temperature and also the packaging medium. However, some components, such as thiamine, are very stable and not noticeably degraded during normal storage. Ascorbic acid, being heat and oxygen sensitive, is easily lost from products stored under aerobic conditions. Peas held for 6 months under a CO_2 atmosphere retained 70% of the original ascorbic acid; but, when these same peas were stored aerobically they only retained 3% (Stevens, 1943). A similar effect was noticed with respect to carotene in carrots, where none was lost under carbon dioxide storage, but 60% was lost under aerobic storage.

In general, niacin, vitamin B_{12}, and also pyridoxine are stable during storage, especially in freeze-dried products (Hollingsworth, 1970); therefore, none of these will be lost. Riboflavin is, however, degraded somewhat during storage.

FORTIFICATION

Nutrition (Nutrification)

Salunkhe and Bolin (1972) fortified apple juice, peach nectar, and cherry juice with solubilized soya protein and then dehydrated the mixes by a

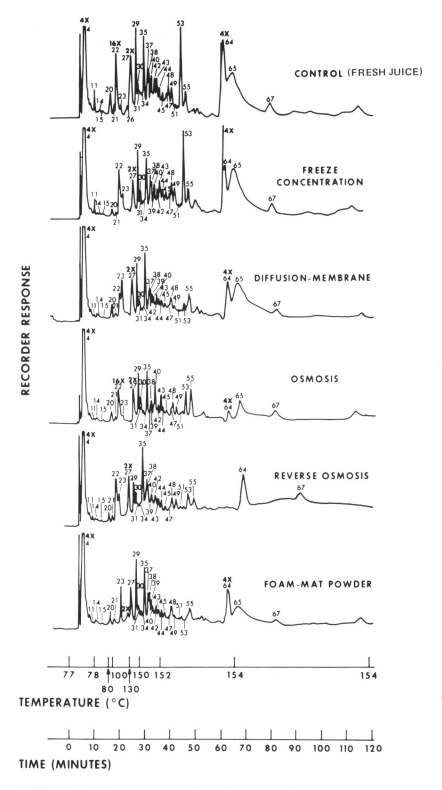

FIGURE 17. GLC chromatograms of volatiles extracted from cherry juice concentrated by various procedures. Note the aroma degradation caused by different dehydration processes. (Source: Bolin and Salunkhe, 1971.)

foam-mat dehydration method to a nutritious powder. These fruit juice powders had a high nutritional content per unit weight of volume which minimizes shipping and storage expenses. Other protein sources could probably be used for such fortification, but each particular protein source, whether fish protein concentrate or a dairy protein, will impart different characteristics to the final reconstituted juice as to viscosity, clarity, etc. This could lead to a variety of protein fortified fruit juice drinks.

Numerous fruit juices, including prune juice, have been fortified with ascorbic acid. An excess of the vitamin must be added, however, to insure that the juice contains the amount declared on the label even after extended storage (Gerber and Salunkhe, 1960). Vitamin C has also been added to dehydrated potato granules, but a pink color develops in this product during storage, evidently because of the oxidation of the ascorbic acid.

Recently, LaChance (1972) proposed a new term, "nutrification," which implies "to make nutritious" and is more meaningful to the consumer. Nutrification is a most rapid, economical, flexible, and socially acceptable method of changing the nutrient intake of a given population. New nutritional regulations by the FDA are proposing to put nutrification labeling on packaged foods, and as a guide to this a new FDA proposed USA Recommended Daily Allowance Table has been issued (FDA, 1973).

Flavor

Dehydrated vegetables lose flavor and develop off flavors when stored at room or higher temperatures. Lee et al. (1953) investigated the feasibility of using vegetable juice concentrates to improve the flavor of dehydrated vegetables. They indicated that concentrates of juices from celery, onion, and turnip can be used in conjunction with dehydrated vegetables to increase palatability. Nury and Watters (1965) made a water extract from ground prune pits and used this extract to enhance the taste of pitted prunes and fresh prune juice.

Texture

The textures of many fruits and vegetables are rather soft, so that improvement in firmness to protect their integrity through handling and processing of dehydration is highly desirable. It is important that the dehydrated products upon rehydration will regain their turgid texture, but not become mushy. Chemical pretreatment of the fruits and vegetables using calcium chloride or alum has been shown to be effective in firming up the texture of the products and has been used in commercial applications. The effect of calcium on plant tissue was pointed out by Kertesz (1939), and he subsequently adapted it to the firming of tomatoes (Kertesz et al., 1940) and apple slices (Kertesz, 1947). The effects of calcium treatment on red tart cherries were studied by Whittenberger (1952), Whittenberger and Hills (1953), Bedford and Robertson (1957), and LaBelle (1971). Improvements of firmness of the treated cherries were demonstrated by these researchers.

Color

Discoloration of the dehydrated fruits and vegetables during storage under high temperatures and high relative humidity is detrimental to the product quality. Erlandson and Wrolstad (1972) reported that the rate of anthocyanin degradation in freeze-dried strawberry puree increased with an increase in relative humidity, at $37^{\circ}C$. The pigments were quite stable at a relative humidity of 11% or below. They also suggested the degradative mechanism was chemical hydrolysis rather than enzymatic. LaJollo et al. (1971) found that chlorophyll degradation in a freeze-dried model system and in freeze-dried blanched puree increased with increasing water activity (a_w). Incorporation of synthetic colors into the dehydrated products during reconstitution may help to improve the color quality of the rehydrated products.

UTILIZATION

General

Dried fruits are utilized mainly in bakery products, in cooked "sauces," and for eating out of the hand. In countries where fresh fruits are not readily available during certain parts of the year, dried fruits are used extensively in their place. Dried foods obviously simplify problems of logistics in military operations due to economies in weight and space. Consequently, the mobilization of modern armies has been accompanied by greatly increased demands for dried fruits and vegetables. For these and other reasons, production of dried or dehydrated fruits and vegetables in the United States is maintained at a high level.

Rehydration

Factors that affect rehydration processes of the dehydrated products are time, temperature, air displacement, pH, and ionic strength (Karel, 1963). Rehydration rates can be accelerated by ultrasonic treatment of the product to be rehydrated in water. Gamma radiation increases the rehydration rates of freeze-dehydrated apples (Lee and Salunkhe, 1966). In addition, it can control microbial growth subsequent to dehydration and during storage. Luh and Eidels (1969) reported that at 26°C freeze-dried mushrooms rapidly reached the maximum rehydratability, while at 98°C the rate of rehydration was slower. The degree of rehydration was also lower at 98°C. Goldblith et al. (1963) also found that the rehydration ratio of freeze-dried shrimp and salmon decreased markedly as the temperature of the rehydration water increased.

New Fabricated Products

Fabricated foods as defined by Glicksman (1971) are "foods designed and built according to plan from individual components, natural or synthetic, to yield products having specified physical (textural), chemical, and functional properties." Fabricated foods prepared from ingredients of dehydrated fruits and vegetables are becoming increasingly popular in today's new product market. Enrichment and fortification are inseparably associated with fabricated foods.

Fruit Drinks

Low moisture fruit disks, dried and compressed into bite-size pieces, have been developed for incorporation in cold breakfast cereals (Nury et al., 1965). Low moisture fruit flakes from ground prunes, apricots, and figs were prepared by modified atmospheric drum dryer. The flakes were then compressed into disks by a pellet press. The bite-size pieces exhibit good fruit flavor, appearance, and stability.

Instant Applesauce

Lazar and Morgan (1966) developed a process for drum-drying pureed apples to 2% moisture applesauce. The finished products (flakes or granules) may be used as thickening agents to improve the rheological characteristics, or as substitutes for insoluble solids derived from other fruits or vegetables in powdered drink and soup mixes, dry fruit sauces, or toppings. In pie mixes, the dry solids act as a natural thickener and anti-boil-out agent. Salunkhe et al. (1971) reported that freeze-dried applesauce has potential from baby food to space feeding because of its high quality and instant reconstitution ability.

Extruded Potato Chips

Uniform shape, size, color, texture, and flavor are the advantages in quality of fabricated (extruded) potato chips when compared to the natural counterpart. The ratio of amylose to amylopectin of the starch being used can be controlled so that the crispness of the extruded chips would not fluctuate from batch to batch. The problem of browning due to reducing sugars in potatoes is eliminated since the amount of sugars in the formulation can be adjusted. Nutritionally it can easily be fortified with protein, vitamins, and minerals.

Puffed Apple Chips

A new apple snack food, apple chips that have the crispy texture of potato chips and a fresh fruit flavor, was developed by Kitson et al. (1972). They employed a two-step process: (1) prefry vacuum infusion treatments of apple slices with various sugar solutions to reduce the fat content of the finished chips, then (2) a subsequent vacuum fry at 98.9 to 104.4°C and 27 to 29 in. vacuum. They found that Maltrin-10, a 10% DE maltro dextrin (Grain Processing Corp., Muscatine, Iowa) effectively reduced fat pickup by the apple chips with a desirable puffing agent. Moisture content of the 0.08 to 0.10 in. thick apple chips is 2% or less.

Soya-banana Powder

Steinberg (1972) described a pilot plant process for making a weaning food using soybean and bananas. Whole soybeans are simultaneously hydrated and blanched for 30 min in boiling water. Ten parts water are next added to one part blanched beans, after which they are then finely ground in a comminuting mill. Fresh ripe bananas are added at a ratio of one part banana solids to one part soybean solids and blended in the same mill with 100 ppm of sodium bisulfite to prevent darkening. The slurry is dehydrated to 3% moisture using a double-drum dryer with 0.01 in. drum spacing and 40 psi steam pressure in the drums. The resulting product has a pleasant flavor, is readily reconstituted, and has a good shelf life. Other kinds of fruits could be used in place of the banana, and other kinds of legumes or oilseeds could be substituted for the soybeans.

Texturized Vegetable Proteins

Striking progress has been made on the utilization of vegetable proteins (oilseed proteins), a unique group of dehydrated vegetable products, in fabricated food manufacturing. Soybeans are the most economical and most practical source of protein today and are used in a multiplicity of ways. New protein beverages are constantly being developed by the simple incorporation of soy flour into various types of drinks. Textured vegetable proteins made by pressure extrusion of the defatted soy flour are increasingly utilized in fabricated foods as meat extender or simulated meat and meat products because of its advantageous meat-like textures. Texturized vegetable protein products are equally nutritious to meat, yet they do not contain fat and cholesterol.

Intermediate Moisture Fruits

Intermediate moisture foods are characterized by a water activity low enough to prevent growth of bacteria, and by conditions minimizing the potential for growth of other microorganisms. Soft and moist dried fruits like prunes and apricots, jams, and some pie fillings belong to this category (Karel, 1973). Intermediate moisture foods offer a number of potential advantages for special military situations. They are concentrated foods which can be eaten without preparation and without the sensation of dryness generally encountered with fully dehydrated products. They are plastic and can be compressed into configurations for maximum packaging and packing efficiency. Their resistance to microbial growth provides an additional margin of safety and wholesomeness (Pavey, 1972). Intermediate moisture fruit bars of 14 variations have been produced from combinations of figs, dates, pears, cherries, almonds, coconuts, oranges, sesames, raisins, and macaroons (Rahman et al., 1971). These fruit bars are made for direct consumption from the package without rehydration and are in the compressed state.

FUTURE RESEARCH

Nonenzymatic browning could be advantageously studied to a greater degree, even though much is already known concerning some of the reactions and reacting intermediates. In fruits, concurrent reactions are occurring between amino acids, sugars, and fruit acids. If any of these reactions could be halted, the overall production of brown pigmented products may be reduced. One area of investigation could be to eliminate the protein element since fruits are not high in protein content. This could possibly be done by tying it up chemically so it would be unavailable to react, but available to the human body.

Sulfur dioxide has long been used as an economically efficient way of retarding enzymatic and nonenzymatic browning. However, in this day and age where any kind of treatment is a suspect from the safety aspect, an alternative procedure is needed that could replace sulfur dioxide if this ever became necessary.

Further nutritional studies would undoubtedly be profitable in relation to determining the effect of each processing step on the nutrient and vitamin losses in fruits and vegetables. Also, the stability of the nutrients in fortified products should be established.

ACKNOWLEDGMENT

The authors express their thanks to Professor M. W. Miller for the critical review of the manuscript. The senior author is especially grateful to his graduate students — Dr. C. Y. Lee, Dr. Y. S. Lee, Dr. W. Sathiraswasti, Dr. A. A. Boe, Dr. F. S. Nury, Dr. J. Y. Do, Dr. H. R. Bolin, Major W. T. Nabors, N. Suthivanit, and C. Srisangnam; and to research associates and collaborators — Professor R. L. LaBelle, Professor C. Brennand, Professor F. Bardwell, Professor F. Taylor, Dr. M. Merkley, Dr. E. Wilcox, G. G. Watters, J. E. Brekke, Dr. A. R. Rahman, Dr. A. S. Henick, Dr. L. H. Pollard, and the late Professor R. K. Gerber for their participation in the research program on dehydration of fruits and vegetables at Utah State University.

REFERENCES

1. **Acker, L.,** Enzymic reactions in foods of low moisture content, *Adv. Food Res.,* 11, 263, 1962.
2. **Allaire, R. P.,** Potential application for the microwave heat exchange, *Food Technol.,* 19, 40, 1965.
3. **Anon.,** *Canner/Packer Yearbook,* Number 131, No. 10, September 25, 1962.
4. **Anon.,** Solvent-drying: New food processing method, *Chem. Eng.,* 60, 1968.
5. **Anon.,** Revolutionary hot snacks, *Food Process. (Chicago),* 33(1), F4, 1972a.
6. **Anon.,** Be patient. Soon—fruit leathers, *Sunset,* July, 1972b.
7. **Anon.,** Dries fruit without sulfur, *Food Process. (Chicago),* 33(5), 25, 1972c.
8. **Anon.,** Keeps powdered materials flowing, *Food Process. (Chicago),* 34(2), 19, 1973a.
9. **Anon.,** Developments to watch, *Food Eng. (N.Y.),* 45(6), 21, 1973b.
10. **Bardwell, H. F., Knowlton, G., and Salunkhe, D. K.,** Food storage in the home, Utah State University Extension Service, Circular 257 (Revised), 1966.
11. **Bardwell, H. F. and Salunkhe, D. K.,** Home drying of fruits and vegetables, Utah State University Extension Service, Circular 332, 1966.
12. **Beattie, J. H. and Gould, H. P.,** Commerical evaporation and drying of fruits, *U.S. Dept. Agric. Farmer's Bull. 9C3,* Washington, D.C., 1917.
13. **Beavens, E. A.,** Cabinet dehydrators suited to small-scale operations, *Food Ind.,* 16(1), 70, 116, 1944a.
14. **Beavens, E. A.,** Cabinet dehydrators suited to small-scale operations, *Food Ind.,* 16(2), 90, 134, 1944b.
15. **Beavens, E. A.,** Cabinet dehydrators suited to small-scale operations, *Food Ind.,* 16(3), 75, 135, 1944c.
16. **Bedford, C. L. and Robertson, W. F.,** Effect of handling and processing methods on the firmness and quality of canned and frozen red cherries, *Mich. Agric. Exp. Stn. Q. Bull.,* 40, 51, 1957.
17. **Bhatia, B. S.,** Dehydration of foods by deep fat frying, *J. Food Sci. Technol. (Mysore),* 7 (suppl.), 14, 1970.
18. **Blau, R., Powell, M., and Gerling, J. E.,** Results of 2450 megacycle microwave treatments in potato chip finishing, *Proceedings of the 28th Annual Conference and Exhibit of the Potato Chip Institute, International,* New York, 1965.
19. **Boggs, M. M and Talburt, W. F.,** Comparison of frozen and dehydrofrozen peas with fresh and stored pod peas, *Food Technol.,* 6, 438, 1952.
20. **Bohrer, B.,** "Azeotropic freeze drying," U.S. Patent No. 3,298,109, 1967.
21. **Bolin, H. R.,** Fruit Juice Concentrates and Powders. I. Development of a New Concentrate Procedure. II. Physicochemical and Volatile Flavor Changes, Ph.D. Dissertation, Utah State University, Logan, Utah, 1970.
22. **Bolin, H. R. and Boyle, F. P.,** Use of potassium sorbate, diethylpyrocarbonate and heat for the preservation of prunes at high moisture levels, *J. Sci. Food Agric.,* 18, 289, 1967.
23. **Bolin, H. R. and Boyle, F. P.,** Effect of storage and processing on sulfur dioxide in preserved fruit, *Food Prod. Dev.,* 6(7), 82, 1972.
24. **Bolin, H. R. and Salunkhe, D. K.,** New ingredients from cherries, *Food Eng. (N.Y.),* 41(8), 44, 1969.
25. **Bolin, H. R. and Salunkhe, D. K.,** Physicochemical and volatile flavor changes occurring in fruit juices during concentration and foam-mat drying, *J. Food Sci.,* 36, 665, 1971.
26. **Bolin, H. R., King, A. D., Jr., Stanley, W. L., and Jurd, L.,** Antimicrobial protection of moisturized Deglet Noar dates, *Appl. Microbiol.,* 23(4), 799, 1972.
27. **Bolin, H. R., Fuller, G., and Powers, J.,** Apricot concentrate product development, *Food Prod. Dev.,* 7(2), 30, 1973.
28. **Bolin, H. R., Nury, F. S., and Bloch, F.,** Effect of light on processed dried fruits, *Food Technol.,* 18(12), 151, 1964.
29. **Bomben, J. L., Dietrich, W. C., Farkas, D. F., Hudson, J. S., DeMarchena, E. S., and Sanshuk, D. W.,** Pilot plant evaluation of individual quick blanching (IQB) for vegetables, *J. Food Sci.,* 38, 590, 1973.
30. **Brekke, J. E. and Nury, F. S.,** Fruits, *Food Dehydration,* Vol. II, VanArsdel, W. B. and Copley, M. J., Eds., Avi Publishing, Westport, Conn., 1964.
31. **Brockmann, M. C.,** Development of intermediate moisture foods for military use, *Food Technol.,* 24, 896, 1970.
32. **Brockmann, M. C.,** Personal communication, 1973.
33. **Brown, G. E., Farkas, D. F., and DeMarchena, E. S.,** Centrifugal fluidized bed blanches, dries, and puffs piece-form foods, *Food Technol.,* 28(12), 23, 1972.
34. **Brynke, C. and Smithies, W. R.,** Vacuum drying, U.S. Patent No. 2,930,139, 1960.
35. **Calloway, D. H.,** Dehydrated foods, *Nutr. Rev.,* 20(9), 257, 1962.
36. **Chiang, J. C., Singh, B., and Salunkhe, D. K.,** Effect of water quality on canned carrots, sweet cherries, and apricots, *J. Am. Soc. Hortic. Sci.,* 96, 1971.
37. **Collins, J. L. and Ruch, B.,** Breaded, deep-fried vegetable soybeans, *Food Prod. Dev.,* 3(6), 40, 1969.
38. **Collins, J. L. and Sanders, G. G.,** Deep-fried snack food prepared from soybeans and onions, *Food Technol.,* 27(5), 46, 1973.
39. **Collins, J. L., Baker, C., and McCarty, I. E.,** Deep-fried corn-on-the-cob, *Tenn. Farm Home Sci. Prog. Rep.,* 78, 8, 1971.
40. **Cooley, A. M., Seversen, D. E., Peightal, D. E., and Wagner, J. R.,** Studies on dehydrated potato granules, *Food Technol.,* 8, 263, 1954.

41. **Copley, M. J., Kaufman, V. F., and Rasmussen, C. L.,** Recent development in fruit and vegetable powder technology, *Food Technol.,* 10,,589, 1956.

42. **Copson, D. A.,** *Microwave Heating,* Avi Publishing, Westport, Conn., 1962.

43. **Cording, J., Jr., Willard, M. J., Jr., Eskew, R. K., and Sullivan, J. F.,** Advances in the dehydration of mashed potatoes, *Food Technol.,* 11, 236, 1957.

44. **Coston, S. and Smith, D. B.,** *Freeze Drying of Foodstuffs,* Columbia Press, Manchester, England, 1962.

45. **Cruess, W. V.,** Evaporators for prune drying, *Calif. Agric. Exp. Stn. Bull.,* Circular 213, 1919.

46. **Cruess, W. V., Christie, A. W., and Flossfeder, F. H. C.,** The evaporation of grapes, *Calif. Agric. Exp. Stn. Bull.,* 322, 1920.

47. **Cruess, W. V.,** *Commercial Fruit and Vegetable Products,* 4th ed., McGraw-Hill, New York, 1958.

48. **Davis, C. O., Smith, O., and Olander, J.,** Microwave processing of potato chips, Pt. I., *Potato Chipper,* 25(2), 38, 1965.

49. **Davis, C. O., Smith, O., and Olander, J.,** Microwave processing of potato chips, Pt. II., *Potato Chipper,* 25(3), 72, 1965.

50. **Davis, C. O., Smith, O., and Olander, J.,** Microwave processing of potato chips, Pt. III., *Potato Chipper,* 25(4), 78, 1965.

51. **DeGroot, A. P.,** The influence of dehydration of foods on the digestibility and the biological value of the protein, *Food Technol.,* 17, 339, 1963.

52. **Do, J. Y. and Salunkhe, D. K.,** Effects of glycerol, propylene glycol, and ethanol on quality of freeze dehydrated slices of celery petioles, *33rd Annual Institute of Food Technologists Meeting,* Miami Beach, Fla., June 10, 1973.

53. **Do, J. Y., Salunkhe, D. K., and Srisangnam, C.,** Effects of dehydration methods on quality and composition of pitted sour cherries, unpublished data, 1973.

54. **Eidt, C. C.,** Principles and methods involved in dehydration of apples, *Can. Dept. Agric., Tech. Bull.,* 18, Publica 625, Ottawa, Canada, 1938.

55. **Eisenhardt, N. H., Cording, J., Jr., Eskew, R. K., and Sullivan, J. F.,** Quick-cooking dehydrated vegetable pieces. I. Properties of potato and carrot products, *Food Technol.,* 16, 143, 1962.

56. **Erlandson, J. A. and Wrolstad, R. E.,** Degradation of anthocyanins at limited water concentration, *J. Food Sci.,* 37, 592, 1972.

57. **Farkas, D. F., Lazar, M. E., and Butterworth, T. A.,** The centrifugal fluidized bed. I. Flow and pressure drop relationships, *Food Technol.,* 23(11), 125, 1969.

58. **Feinberg, B., Schwimmer, S., Reeve, R., and Juilly, M.,** Vegetables, *Food Dehydration,* Vol. II, Van Arsdel, W. B. and Copley, M. J., Eds., Avi Publishing, Westport, Conn., 1964.

59. **Feustel, I. C., Hendel, C. I., and Juilly, M. E.,** Potatoes, *Food Dehydration,* Vol. II, Van Arsdel, W. B. and Copley, M. J., Eds., Avi Publishing, Westport, Conn., 1964.

60. **F.D.A.,** Food labeling, Federal Register, 38(13), 2124, 1973.

61. **Gelashvili, E. D.,** Freezing of figs, *Konservnaya i Ovoshchesushilnaya Promyshlennost',* 26(12), 17, 1971.

62. **Gentry, J. P., Miller, M. W., and Claypool, L. L.,** Engineering and fruit quality aspects of prune dehydration in parallel and counter-flow tunnels, *Food Technol.,* 19(9), 1427, 1965.

63. **Gerber, R. K. and Salunkhe, D. K.,** Vitamin enriched apple juice, *Utah Science,* 21(2), 46 and 63, 1960.

64. **Glicksman, M.,** Fabricated foods, *CRC Crit. Rev. Food Technol.,* 2(1), 21, 1971.

65. **Goldblith, S. A., Karel, M., and Lusk, G.,** The role of food science and technology in the freeze dehydration of foods, *Food Technol.,* 17, 139, 1963.

66. **Gooding, E. G. B. and Rolfe, E. J.,** Some recent work on dehydration in the United Kingdom, *Food Technol.,* 11, 306, 1957.

67. **Gould, H. P.,** Evaporation of apples, *U.S. Dept. Agric. Farmers' Bull.,* 291, Washington, D. C., 1907.

68. **Graham, R. P., Huxsoll, C. C., Hart, M. R., Weaver, M. L., and Morgan, A. I., Jr.,** "Dry" caustic peeling of potatoes, *Food Technol.,* 23(2), 61, 1969.

69. **Hanson, S. W. F.,** The accelerated freeze drying (AFD) method for food preservation, Ministry of Fisheries and Food, London, 1961.

70. **Hayes, N. V., Cotton, R. H., and Roy, W. R.,** Problems in the dehydration of orange juice, *Proc. Am. Soc. Hortic. Sci.,* 47, 123, 1946.

71. **Hearne, J. F. and Tapsfield, D.,** Some effects of reducing, during storage, the water content of dehydrated strip potato, *J. Sci. Food Agric.,* 7(11), 210, 1956.

72. **Hendel, C. E., Burr, H. K., and Boggs, M. M.,** Nonenzymatic browning and sulfite flavor in dehydrated white potato. Effects of sulfite level and method of sulfite application, *Food Technol.,* 9, 627, 1955.

73. **Heyman, W. A.,** Porous expanded citrus fruit products, U.S. Patent No. 2,328,554, 1943.

74. **Hodge, J. E.,** Chemistry of browning reactions in model systems, *J. Agric. Food Chem.,* 1, 928, 1953.

75. **Holdsworth, S. D.,** Dehydration of food products, *J. Food Technol.,* 6, 331, 1971.

76. **Hollingsworth, D. F.,** Effects of some new production and processing methods on nutritive values, *J. Am. Diet. Assoc.,* 57(3), 247, 1970.

77. **Hoover, M. W. and Miller, N. C.,** Process for producing sweet-potato chips, *Food Technol.,* 27(5), 74, 1973.

78. **Hope, G. W. and Vitale, D. G.,** Osmotic dehydration. International Development Research Centre Monograph IDRC-004e, 12 pp., Canada Dept. of Agric., Food Res. Inst., Ottawa, Canada, 1972.

79. **Huxsoll, C. C. and Morgan, A. I., Jr.,** Use of microwaves in the food industry, *26th Institute of Food Technologists Annual Meeting,* Portland, Ore., May 22, 1966.

80. **Huxsoll, C. C., Dietrich, W. C., and Morgan, A. I., Jr.,** Comparison of microwave with steam or water blanching of corn-on-the-cob, *Food Technol.,* 47(3), 84, 1970.

81. **Jeppson, M. R.,** Techniques of continuous microwave food processing, *Cornell Hotel and Restaurant Admin. Q.,* 5(1), 60, 1964.

82. **Karel, M.,** Physical and chemical considerations in freeze dehydrated foods, in *Exploration in Future Food-processing Techniques,* Goldblith, S. A., Ed., The M.I.T. Press, Cambridge, Mass., 1963.

83. **Karel, M. and Nickerson, J. T. R.,** Effects of relative humidity, air, and vacuum on browning of dehydrated orange juice, *Food Technol.,* 18, 104, 1964.

84. **Karel M.,** Recent research and development in the field of low-moisture and intermediate moisture foods, *CRC Crit. Rev. Food Technol.,* 3(3), 329, 1973.

85. **Kaufman, V. F.,** Costs and methods for piestock apples, *Food Eng. (N.Y.),* 23, 97, 1951.

86. **Kelley, E. G. and Baum, R. R.,** Preparation of tasty vegetable products by deep-fat frying, *Food Technol.,* 9, 388, 1955.

87. **Kertesz, Z. I.,** The effect of calcium in plant tissues, *Canner,* 88, 26, 1939.

88. **Kertesz, Z. I.,** Calcium treatment improves pie apples, *Food Packer,* 28, 30, 1947.

89. **Kertesz, Z. I., Tolman, T. G., Loconti, J. D., and Ruyle, E. H.,** The use of calcium in the commercial canning of whole tomatoes, *N.Y. Agric Exp. Stn. Tech. Bull.,* 252, 1940.

90. **Kilpatrick, P. W., Lowe, E., and Van Arsdale, W. B.,** Tunnel dehydrators for fruits and vegetables, *Adv. Food Res.,* 6, 314, 1955.

91. **Kitson, J. A., Lackey, C. L., and Wright, H. T.,** New fruit flavor snack, *Food Eng. (N.Y.),* 44(11), 50, 1972.

92. **Kitson, J. A., Lackey, C. L., and Coltart, M. L.,** Dry sauces, soup mixes reap benefits of new ingredient: low moisture apple solids, *Food Prod. Dev.,* 6(7), 36, 1972.

93. **Kohman, S.,** The preservation of the nutritive value of foods in processing, *J.A.M.A.,* 120, 831, 1942.

94. **LaBelle, R. L.,** Reconstitution of dried foods, *Proceedings of Symposium on Frontiers in Food Research,* Cornell Univ., Ithaca, N.Y. April 12, 1966.

95. **LaBelle, R. L.,** Heat and calcium treatments for firming red tart cherries in a hot-fill process, *J. Food Sci.,* 36, 323, 1971.

96. **LaBelle, R. L. and Moyer, J. C.,** Dehydrofreezing of red tart cherries, *Food Technol.,* 20, 1345, 1966.

97. **Labuza, T. P.,** Symposium 2: Effects of dehydration and storage, *Food Technol.,* 27(1), 20, 51, 1973.

98. **LaChance, P. A.,** Nutrification: A concept assuring nutritional quality by primary intervention in feeding system, *Food Technol.,* 20, 522, 1972.

99. **LaJollo, F., Tannenbaum, S. R., and Labuza, T. P.,** Reaction at limited water concentration. 2. Chlorophyll degradation, *J. Food Sci.,* 36, 850, 1971.

100. **Langworthy, C. F.,** Raisins, figs, and other dried fruits, and their use, *Yearb. U.S. Dept. Agric.,* Washington, D.C., 1912.

101. **Lawhorn, J. T., Cater, C. M., and Mattil, K. F.,** Preparation of a high-protein low-cost nut-like food product from glandless cottonseed kernels, *Food Technol.,* 24, 701, 1970.

102. **Lazar, M. E.,** Dehydrofreezing of fruits and vegetables, *The Freezing Preservation of Foods,* 4th ed., Vol. 3, Avi Publishing, Westport, Conn., 1968.

103. **Lazar, M. E. and Farkas, D. F.,** The centrifugal fluidized bed. 2. Drying studies on piece-form foods, *J. Food Sci.,* 36, 315, 1971.

104. **Lazar, M. E. and Morgan, A. I., Jr.,** Instant applesauce, *Food Technol.,* 20(4), 179, 1966.

105. **Lazar, M. E. and Miers, J. C.,** Improved drum-dried tomato flakes are produced by a modified drum drier, *Food Technol.,* 25, 830, 1971.

106. **Lazar, M. E., Chapin, E. O., and Smith, G. S.,** Dehydrofrozen apples: Recent developments in processing methods, *Food Technol.,* 15(1), 32, 1961.

107. **Lazar, M. E., Lund, D. B., and Dietrich, W. C.,** A new concept in blanching. IQB reduces pollution while improving nutritive value and texture of processed foods, *Food Technol.,* 25(7), 24, 1971.

108. **Lazar, M. E., Barta, J. E., and Smith, G. S.,** Dry-blanch-dry (DBD) method for drying fruit, *Food Technol.,* 17(9), 120, 1963.

109. **Lazar, M. E., Smith, G. S., and Chapin, E. O.,** Preparation of dehydrated fruit having the characteristics of sun-dried fruit, U.S. Patent No. 2,979,412, 1961.

110. **Lee, C. Y. and Salunkhe, D. K.,** Effects of gamma radiation on freeze-dehydrated apples (*Pyrus malus*), *Nature,* 210(5039), 971, 1966.

111. **Lee, C. Y. and Salunkhe, D. K.,** Effects of dehydration processes on color and rehydration of fruits, *J. Sci. Food Agric.,* 18, 566, 1967.

112. **Lee, C. Y. and Salunkhe, D. K.,** Sucrose penetration in osmo-freeze dehydrated apple slices, *Curr. Sci.*, 37(10), 297, 1968.

113. **Lee, C. Y., Salunkhe, D. K., and Nury, F. S.,** Effects of dehydration processes on flavor compounds and histology of apricots (*Prunus armeniaca*), *J. Sci. Food Agric.*, 17, 393, 1966.

114. **Lee, C. Y., Salunkhe, D. K., Watters, G. G., and Nury, F. S.,** Effects of dehydration processes on flavor compounds and histology of peaches (*Prunus persica*), *Food Technol.*, 20, 141, 1966.

115. **Lee, C. Y., Salunkhe, D. K., and Nury, F. S.,** Some chemical and histological changes in dehydrated apples, *J. Sci. Food. Agric.*, 8, 89, 1967.

116. **Lee, F. A., Hening, J. C., and Pederson, C. S.,** Improvement of the flavor of dehydrated vegetables by the addition of vegetable juice concentrates, *Food Technol.*, 7(4), 162, 1953.

117. **Lewin, L. M., and Mateles, R. I.,** Freeze drying without vacuum: A preliminary investigation, *Food Technol.*, 16, 94, 1962.

118. **Lowe, E., Ramage, W. D., Durkee, E. L., and Hamilton, W. R.,** Belt trough, a new continuous dehydrator, *Food Eng. (N.Y.)*, 24, 43, 1955.

119. **Luh, B. S. and Eidels, L.,** Chemical changes in freeze-dried mushrooms (*Agaricus bisporus*), *Confructa*, 14, 8, 1969.

120. **Malecki, G. J., Shinde, P., Morgan, A. I., Jr., and Farkas, D. F.,** Atmospheric fluidized bed freezing, *Food Technol.*, 24, 601, 1970.

121. **McBean, D. McG., Miller, M. W., Johnson, A. A., and Pitt, J. I.,** Sulphur dioxide levels for sulphuring tree fruits before drying, *C.S.I.R.O. Food Preserv. Q.*, 27(1), 23, 1967.

122. **McLain, L. P. and Abott, J. C.,** Cryogenic advances in food technology, *Process Biochem.*, 6(10), 9, 13, 1971.

123. **Medriczky, A.,** A bright future for potato dehydration is warranted, *Am. Potato J.*, 37, 77, 1960.

124. **Merson, R. L. and Morgan, A. I., Jr.,** Juice concentration by reverse osmosis, *Food Technol.*, 22, 97, 1968.

125. **Miller, M. W.,** The drying of fruits in Australia and California, *C.S.I.R.O. Food Preserv. Q.*, 25(2), 33, 1965.

126. **Miller, M. W., Fridley, R. B., and McKillop, A. A.,** The effect of mechanical harvesting on the quality of prunes, *Food Technol.*, 17(11), 1451, 1963.

127. **Molaison, L. J., Spadaro, J. J., Roby, M. T., and Lee, F. H.,** Dehydrated diced sweet potatoes – a pilot plant process and production evaluation, *Food Technol.*, 16(11), 101, 1962.

128. **Morgan, A. F., and Field, A.,** The effect of drying and of sulfur dioxide upon the antiscorbutic property of fruit, *J. Biol. Chem.*, 82, 579, 1929.

129. **Morgan, A. F. and Field, A.,** Vitamins in dried fruits. II. The effects of drying and of sulfur dioxide upon the vitamin A content of fruits, *J. Biol. Chem.*, 88, 9, 1930.

130. **Morgan, A. F., Kimmell, L., Field, A., and Nichols, P. F.,** The vitamin content of sultanina (Thompson seedless) grapes and raisins, *J. Nutr.*, 9, 369, 1935.

131. **Morgan, A. I., Jr., Graham, R. P., Ginnette, L. F., and Williams, G. S.,** Recent developments in foam-mat drying, *Food Technol.*, 15, 37, 1961.

132. **Morgan, A. I., Jr., Lowe, E., Merson, R. L., and Durkee, E. L.,** Reverse osmosis, *Food Technol.*, 19, 52, 1965.

133. **Moyer, J. C.,** The nutritive value of dehydrated vegetables, *J. Am. Diet. Assoc.*, 19, 14, 1943.

134. **Mrak, E. M. and Phaff, H. J.,** Sun drying of fruits, *Calif. Agric. Exp. Stn. Circ.*, 392, 1949.

135. **Mrak, E. M., Fisher, C. D., and Bornstein, B.,** The effect of certain substances and pretreatment on the retention of color and sulfur dioxide by dried cut fruit, *Fruit Prod. J.*, 21(10), 297, 1942.

136. **Nabors, W. T. and Salunkhe, D. K.,** Pre-fermentation inoculations with *Leuconostoc mesenteroides* and *Lactobacillus plantarum* on physicochemical properties of fresh and dehydrated sauerkraut, *Food Technol.*, 23(3), 67, 1969.

137. **Nakamura, T.,** British Patent No. 1,093,598, 1967.

138. **Neel, G. H., Smith, G. S., Cole, M. W., Olson, R. L., Harrington, W. O., and Mullins, W. R.,** Drying problems in the add-back process for production of potato granules, *Food Technol.*, 8, 230, 1954.

139. **Nichols, P. F. and Christie, A. W.,** Drying cut fruits, *Calif. Agric. Exp. Stn. Bull.*, 485, 1930.

140. **Nonaka, M., Weaver, M. L., and Hautala, E.,** Texturizing process controls crispness and rigidity of French fried potatoes, *Food Technol.*, 26(4), 61, 1972.

141. **Nury, F. S. and Salunkhe, D. K.,** Effects of microwave dehydration on components of apples, *U.S. Dept. Agric. ARS Spec. Bull.*, 74-45, 1968.

142. **Nury, F. S. and Watters, G. G.,** Process for fortifying the flavor of prune juice, U.S. Patent No. 3,211,557, 1965.

143. **Nury, F. S., Bolin, H. R., and Lazar, M. E.,** Dried fruit disks for cold cereals, *Food Eng. (N.Y.)*, 37(8), 85, 1965.

144. **Nury, F. S., Miller, M. W., and Brekke, J. E.,** Preservative effect of some antimicrobial agents on high-moisture dried fruits, *Food Technol.*, 14, 113, 1960.

145. **Nury, F. S., Taylor, D. H., and Brekke, J. E.,** Research for better quality in dried fruits. Figs, *U.S. Dept. Agric. ARS 74-16*, 1960.

146. **Oldencamp, H. A. and Small, R. F.,** Freeze drying apparatus with inflatable platen contact heating, U.S. Patent No. 3,199,217, 1965.

147. **O'Meara, J. P.,** Freeze drying: With or without vacuum, *Food Eng. (N.Y.)*, 35, 55, 1963.

148. **O'Meara, J. P.,** Progress report on microwave drying, *Proceedings of the 29th Annual Conference and Exhibit of the Potato Chip Institute, International,* Las Vegas, Nevada, 1966.

149. **Pavey, R.,** Controlling the amount of internal aqueous solution in intermediate moisture foods. Technical Report 73-17-FL, *U.S. Army Natick Lab. Tech. Rep.,* 1972.
150. **Perry, R. L., Mrak, E. M., Phaff, H. J., Marsh, G. L., and Fisher, C. D.,** Fruit dehydration. I. Principle and equipment, *Calif. Agric. Exp. Stn. Bull.,* 698, 1946.
151. **Pollack, G. A. and Foin, L. C.,** Comparative heating efficiency of microwave and conventional electric oven, *Food Technol.,* 14, 454, 1960.
152. **Ponting, J. D. and McBean, D. M.,** Temperature and dipping treatment effects on drying rates and drying times of grapes, prunes, and other waxy fruits, *Food Technol.,* 24, 85, 1970.
153. **Ponting, J. D., Watters, G. G., Forrey, R. R., Jackson, R., Stanley, W. L., and Robe, K.,** More flavorful dried fruit, *Food Process. (Chicago),* February, 1966a.
154. **Ponting, J. D., Watters, G. G., Forrey, R. R., Jackson, R., and Stanley, W. L.,** Osmotic dehydration of fruits, *Food Technol.,* 20, 125, 1966b.
155. **Porter, V. L., Nelson, A. L., Steinberg, M. P., and Wei, L. S.,** Microwave finish drying of potato chips, *J. Food Sci.,* 38, 583, 1973.
156. **Possingham, J. V.,** Surface wax structure in fresh and dried Sultana grapes, *Ann. Bot. (Lond.),* 36, 933, 1972.
157. **Prescott, S. C.,** Relation of dehydration to agriculture, *U.S. Dept. Agric.,* Office of Secretary, Circular 126, Washington, D.C., 1919.
158. **Prescott, S. C. and Proctor, B. E.,** *Food Technology,* McGraw-Hill, New York, 1937.
159. **Pugsley, C. W.,** A successful community drying plant, *U.S. Dept. Agric. Farmers' Bull.,* 916, Washington, D.C., 1917.
160. **Rahman, A. R., Bishov, S., and Westcott, D. E.,** Reversible compression of dehydrated peas, *J. Texture Stud.,* 2, 240, 1971.
161. **Rahman, A. R., Schafer, G., Prell, P., and Westcott, D. E.,** Non-reversible compression of intermediate moisture fruit bars, Technical Report 71-60-FL, *U.S. Army Natick Lab. Tech. Rep.,* 1971.
162. **Rahman, A. R., Schafer, G., Taylor, G. R., and Westcott, D. E.,** Studies on reversible compression of dehydrated vegetables. Technical Report 70-36-FL, *U.S. Army Natick Lab. Tech. Rep.,* 1969.
163. **Rahman, A. R., Taylor, G. R., Schafer, G., and Westcott, D. E.,** Studies of reversible compression of freeze dried RTP cherries and blueberries. Technical Report 70-52-FL, *U.S. Army Natick Lab. Tech. Rep.,* 1970.
164. **Ralls, J. W., Maagdenberg, J., Yacoub, N. L., Hamnick, D., Zinnecker, M., and Mercer, W. A.,** In-plant, continuous hot-gas blanching of spinach, *J. Food Sci.,* 38(2), 192, 1973.
165. **Rasmussen, C. L., Venstrom, D. W., Neumann, H. J., Olsen, R. L., Rockwell, W. C., and Strong, K.,** Latest dehydro-frozens cut costs, *Food Eng. (N.Y.),* 29, 117, 1957.
166. **Reddy, P. J., Collins, J. L., McCarty, I. E., and Johnston, M. R.,** Sensory evaluation of breaded, deep-fried turnip slices, *Food Prod. Dev.,* 4(8), 38, 1971.
167. **Roberts, A. C., and McWeeny, D. J.,** The use of sulfur dioxide in the food industry, a review, *J. Food Technol.,* 7, 221, 1972.
168. **Rockland, L. B.,** Water activity and storage stability, *Food Technol.,* 23(10), 11, 1969.
169. **Rushton, E., Stanley, E. C., and Scott, A. W.,** Compressed dehydrated vegetable blocks, the application of high frequency heating, *Chem. Ind. (Lond.),* 35, 274, 1945.
170. **Salem, S. A. and Hegazi, S. M.,** Chemical changes occurring during the processing of sun-dried apricot juice, *J. Sci. Food Agric.,* 24, 123, 1973.
171. **Salunkhe, D. K.,** Preparation, preservation, and quality of deep-fat fried products of fruits and vegetables, *Food Technol.,* 10(5), 22, 1956.
172. **Salunkhe, D. K.,** Food Dehydration: Past, Present, and Future, A paper presented at the Moscow Technological Institute for Food Industry, Moscow, U.S.S.R., 1967.
173. **Salunkhe, D. K. and Bolin, H. R.,** Dehydrated protein-fortified fruit juices, *Food Prod. Dev.,* 6(1), 84, 1972.
174. **Salunkhe, D. K., Bolin, H. R., and Suthivanit, N.,** Production and quality of cherry raisins and pickles, *Utah Sci.,* 29(4), 99, 1968.
175. **Salunkhe, D. K., Brennand, C. P., and Bolin, H. R.,** Dehydrated fruits and their utilization, *Utah Sci.,* 32(4), 123, 1971.
176. **Salunkhe, D. K., Chiang, J., and Singh, B.,** Water quality determines quality of canned products, *Utah Sci.,* 32(1), 18, 1971.
177. **Salunkhe, D. K., Lee, Y. S., Bolin, H. R., and Rahman, A. R.,** Quality evaluation of canned and freeze-dried applesauce, *Food Prod. Dev.,* 5(5), 70, 1971.
178. **Salunkhe, D. K., Lee, Y. S., and Bolin, H. R.,** Freeze-dehydrated applesauce has potential, from baby food to space feeding, *Utah Sci.,* 32(1), 15, 1971.
179. **Salunkhe, D. K., Watters, G. G., Taylor, D. H., and Do, J. Y.,** Fruit juice powders, *Utah Sci.,* 24, 56, 74, 1963.
180. **Sapers, G. M., Panasiuk, O., Talley, F. B., and Shaw, R. L.,** Flavor quality and stability of potato flakes. Effects of raw material and processing, *J. Food Sci.,* 38, 586, 1973.
181. **Sathiraswasti, W. and Salunkhe, D. K.,** Quality evaluation of deep-fat fried peas, *Food Technol.,* 12(7), 351, 1958.
182. **Schrader, A. L. and Thompson, A. H.,** Factors influencing the keeping quality of dehydrated apples, *Proc. Am. Soc. Hortic. Sci.,* 49, 125, 1947.

183. **Schroeder, A. L. and Cotton, R. H.,** Dehydration of orange juice, *Ind. Eng. Chem.,* 40, 803, 1948.

184. **Schroeder, H. A.,** Losses of vitamins and trace minerals resulting from processing and preservation of foods, *Am. J. Clin. Nutr.,* 24, 562, 1971.

185. **Schwarz, H. W.,** Dehydration of heat-sensitive materials, *Ind. Eng. Chem.,* 40, 2028, 1948.

186. **Schwarz, H. W. and Penn, F. E.,** Production of orange juice concentrate and powder, *Ind. Eng. Chem.,* 40, 938, 1948.

187. **Seltzer, E. and Settlemeyer, J. T.,** Spray drying of foods, *Adv. Food Res.,* 2, 399, 1949.

188. **Shipman, J. W., Rahman, A. R., Segars, R. A., Kapsalis, J. G., and Westcott, D. E.,** Improvement on the texture of dehydrated celery by glycerol treatment, *J. Food Sci.,* 37(4), 568, 1972.

189. **Simon, M., Wagner, J. R., Silveira, V. G., and Hendel, C. E.,** Calcium chloride as a non-enzymatic browning retardant for dehydrated white potatoes, *Food Technol.,* 9, 271, 1955.

190. **Sluder, J. C., Olsen, R. W., and Kenyon, E. M.,** A method for the production of dry powdered orange juice, *Food Technol.,* 1, 85, 1947.

191. **Smithies, W. R. and Blakeley, T. S.,** Design of freeze-drying equipment for the dehydration of foodstuffs, *Food Technol.,* 13, 610, 1959.

192. **Song, P. and Chichester, C. O.,** Kinetic behavior and mechanism of inhibition in the Maillard reaction. IV. Mechanism of the inhibition, *J. Food Sci.,* 32, 107, 1967.

193. **Stadtman, E. R.,** Nonenzymatic browning in fruit products, *Adv. Food Res.,* 1, 325, 1948.

194. **Stafford, A. E. and Bolin, H. R.,** Improves fruit resulfuring, *Food Eng. (N.Y.),* 44(11), 128, 1972.

195. **Stafford, A. E., Fuller, G., Bolin, H. R., and Mackey, B.,** Fatty acid ester analysis of treated raisins, in press, 1973.

196. **Steinberg, M.,** Food products from whole soybeans, *Soybean Dig.,* 31(3), 32, 1972.

197. **Stevens, H. P.,** Preliminary study of conditions affecting the nutritive values of dehydrated vegetables, *J. Am. Diet. Assoc.,* 19, 832, 1943.

198. **Strashun, S. I. and Talburt, W. F.,** Stabilized orange juice powder. I.

198. **Strashun, S. I. and Talburt, W. F.,** Stabilized orange juice powder. I. Preparation and packaging, *Food Technol.,* 8, 40, 1954.

199. **Strashun, S. I. and Talburt, W. F.,** Puffed powder from juice, *Food Eng., (N.Y.),* 25(3), 59, 1953.

200. **Thijssen, H. A. C.,** Concentration processes for liquid foods containing volatile flavors and aromas, *J. Food Technol.,* 5(3), 221, 1970.

201. **Thomas, M. and Calloway, D.,** Nutritional value of dehydrated food, *J. Am. Diet. Assoc.,* 39, 105, 1961.

202. **Thompson, J. E.,** Process for dehydrating meat, U.S. Patent No. 3,211,599, 1965.

203. **Trauberman, L.,** Processing advances from abroad, *Food Eng. (N.Y.),* 45, 84, 1973.

204. **Tressler, D. K.,** Nutritive value of dried and dehydrated fruits and vegetables, *N.Y. Agric. Exp. Stn. Technol. Bull.,* 262, 1942.

205. **Tressler, D. K.,** New developments in dehydration of fruits and vegetables, *Food Technol.,* 10, 119, 1956.

206. **Tuomy, J. M.,** Freeze drying of foods for the armed services. Technical Report 70-43-FL, *U.S. Army Natick Lab. Tech. Rep.,* 1970.

207. **U.S. Dept. Agric.,** Vegetable and fruit dehydration, a manual for plant operators, Miscellaneous Publication 540, Washington, D.C., 1944.

208. **U.S. Dept. Agric.,** Management handbook to aid emergency extension of dehydration facilities for vegetables and fruits, Western Utilization Research and Development Division, ARS, Albany, Calif., 1959.

209. **Van Arsdel, W. B.,** Principles of the drying process, with special reference to vegetable dehydration. *U.S. Dep. Agric., Bureau Agric. Ind. Chem.,* AID-300, 1951a.

210. **Van Arsdel, W. B.,** Tunnel and truck dehydrators, as used in vegetable dehydration, U.S.D.A., Bur. Agr. Ind. Chem. AIC-308, 1951b.

211. **Van Arsdel, W. B. and Copley, M. J.,** *Food Dehydration, Vol. II,* Avi Publishing, Westport, Conn., 1964.

212. **Van Arsdel, W. B. and Copley, M. J.,** *Food Dehydration, Vol. I,* Avi Publishing, Westport, Conn., 1963.

213. **Von Loesecke, H. W.,** *Drying and Dehydration of Foods,* Reinhold, New York, 1955.

214. **Webster, R. C. and Benson, E. J.,** Preservation of tomatoes by freezing, U.S. Patent No. 3,250,630, 1966.

215. **Weigand, E. H.,** Recirculation driers, *Oreg. Agric. Exp. Stn. Tech. Bull.,* Circulation 40, 1923.

216. **Whittenberger, R. T.,** Factors which affect the drained weight and other characteristics of heat processed red cherries, *Food Res.,* 17, 299, 1952.

217. **Whittenberger, R. T. and Hills, C. H.,** Studies on the cooking of red cherries caused by bruising, cooking, and soaking, *Food Technol.,* 7, 29, 1953.

218. **Wolford, E. R., Ingalsbe, D. W., and Boyle, F. P.,** Freezing of peaches and sweet cherries in liquid N and in dichlorodifluoromethane and behavior upon thawing of strawberries and raspberries, *Proceedings of the 12th International Congress of Refrigeration,* 3, 459, 1967.

219. **Wolford, E. R., Jackson, R., and Boyle, F. P.,** Quality evaluation of stone fruits and berries frozen in liquid nitrogen and in Freezant-12, *Chemical Engineering Program Symposium Series,* 67(108), 131, 1971.

220. **Wong, F. F., Dietrich, W. C., Harris, J. G., and Lindquist, F. E.,** Effect of temperature and moisture on storage stability of vacuum-dried tomato juice powder, *Food Technol.,* 10(2), 96, 1956.

221. **Woodward, T. H.,** Freeze drying without vacuum, *Food Eng. (N.Y.),* 35, 96, 1963.

THE USE OF CERTAIN CHEMICALS TO INCREASE NUTRITIONAL VALUE AND TO EXTEND QUALITY IN ECONOMIC PLANTS

Authors: **M. T. Wu**
D. K. Salunkhe
Department of Nutrition and Food Science
Utah State University
Logan, Utah

INTRODUCTION

The capacity of a civilization to meet its nutritional requirements through adequate food supplies has been a basic determinant of stability and progress. The President's Science Advisory Committee, in their 1967 report on the world food situation, predicted that food output of the underdeveloped world would have to double by the mid-1980's in order to deal with both the population problem and the gross caloric deficit (U.S. Government, 1967).

The most efficient production of dietary calories in tropical and subtropical regions of the world, under their prevailing ecological conditions and economic and technical resources, is achieved by cultivating plant foods of high carbohydrate content. These include cereal grains and starchy roots and fruits. Thus, wheat, rice, corn, cassava, and banana have become staple foods of the majority of the world's population.

The production of protein-rich foods is much less efficient in the tropical and subtropical areas in terms of calories. The protein in the diets of people living in these areas, therefore, is usually very low in concentration and biological value, even though caloric requirements may be satisfied.

When calories are also limited, the protein deficit becomes exaggerated because at least part of the dietary protein must be utilized for energy. Protein deficiency, with or without a caloric deficit, presently constitutes one of the most serious nutritional problems affecting most of the world's population.

This problem could become much more serious in the near future because of the present trend in population growth. The existing trend demands a rapid increase in food production particularly in the underdeveloped countries, precisely where food production is in many instances already inadequate.

Sufficient food of high nutritional value is essential for good health. To combat malnutrition and undernutrition throughout the world, adequate diets containing satisfactory amounts of nutrients are necessary. Since two thirds of the world's population is affected by protein-calorie malnutrition and undernutrition, concentrated efforts must be made to increase both yield and nutritional value of agricultural production.

Malnutrition and undernutrition retard physical growth and mental development, and decrease an individual's resistance to diseases, consequently causing a shorter life. Emotional stability and mental efficiency are lessened and as a result, capacity to work is decreased. These effects are matters of great concern. Severe malnutrition during the first months of life seriously and permanently damages the central nervous system (Scrimshaw and Gordon, 1968). Mild to moderate forms of protein-calorie malnutrition are much more frequent, and usually affect children. A correlation has been demonstrated between the retarded physical growth of these children, which is primarily due to malnutrition, and their performance on behavioral tests, which suggests a similar retardation in mental development.

An increased supply of food is urgently needed to provide adequate calories, protein, vitamins, and minerals by increasing the productivity and efficiency of plant foods. Recently, production has been increased by cultural, genetic, fertilization, and irrigation methods. Control of pests and diseases has increased to some extent the productivity of several crops.

Plant foods are basic in the nutrition of man. A major breakthrough in developing high yielding cultivars of wheat, rice, corn, and sorghum has caused a "Green Revolution." Similar practical and economical breakthroughs to produce high yields of highly nutritious crops seem likely to be associated with the application of low concentrations of certain chemicals.

Increasing yields, however, are pointless if the harvested crops are not consumed by people. It is known that 40% of the produce harvested is never consumed, due to spoilage which takes place from the time of harvest, during storage, transportation, and processing, until it reaches the consumer. Losses due just to postharvest deterioration and diseases of fresh produce currently amount to several billion dollars annually. In the United States about 25% of the produce harvested is never consumed because of spoilage. In developing countries where modern facilities of refrigeration and controlled storage are less common, this figure could be over 60%. Any treatment that could reduce or prevent these losses would help increase the available world food supply.

In this paper we will discuss the effects of certain chemical treatments (when applied in minute quantities to plants and/or to soil) in terms of increased yields, carbohydrates, proteins (quantity and quality), and vitamin and mineral contents. Color changes and the retention of nutrients and quality in the plant product after harvest and during extended storage are also evaluated.

YIELD: QUANTITY AND QUALITY

Chemicals used to increase growth, yield, and nutritional value of economic plants can include fertilizers, herbicides, insecticides, fungicides, and others. Increases in yields are generally due to the elimination of pests and/or competition from other plants. These constitute direct chemical effects. Some chemicals, however, even in minute quantities, can indirectly influence the plant growth and photosynthesis. They may act as hormones to modify the plant's physiological processes, or they may increase the uptake of certain minerals and water, thus altering the nutritional composition of the plants. Both tonnage per acre and ultimate nutritive value should be enhanced.

ESSENTIAL NUTRIENTS FOR THE HUMAN BODY

Proteins

Protein constitutes some 54% of the total organic material of the animal body. For most people in the U.S., some 10 to 14% of the total energy of their diet is contributed by protein. This percentage tends to remain constant irrespective of age, sex, pregnancy, lactation, work, or recovery from illness.

Without adequate protein in their diets, normal growth and health cannot be maintained. One of the most common types of malnutrition, especially in children, is protein deficiency. Evidence is accumulating that a protein deficiency during preschool years can retard mental and physical development. The daily per capita protein intake in developed countries is about 90 g; in many developing countries the figure is 60 g; while in some countries, particularly in South East Asia, it is barely 40 g. In the developing countries protein of vegetable origin accounts for over 80% of all available proteins. Cereals alone provide some 60% and pulses, oilseeds, and nuts about 17%. When other sources of calories are inadequate, some of this protein must be used as a source of energy and this is wasteful.

One solution to the existing protein shortage lies in augmenting the quantity of protein in cereals, legumes, tubers through development of better cultivars, increased use of fertilizers, and related improvements in cultural practices. The most commonly used vegetable or cereal proteins are incomplete in the amino acid patterns as required by man. In general, cereals are deficient in lysine and threonine. It is essential to improve the quality, and especially the balance, of essential amino acids in plant proteins. The improvement of quantity and quality of protein in cereals, legumes, and tubers offers an additional opportunity for achieving maximum impact on protein needs.

Carbohydrates

The food resources of the world are based mainly on staples such as cereals — rice, wheat, corn, sorghum, and barley; legumes — beans, soybeans, and peanuts; root and tuber crops — potatoes, sweet potatoes, and cassava; and fruit crops — banana and coconut. These crops are rich sources of carbohydrates and calories. Since grains, legumes, roots and tubers are being, or could be, produced in large quantities at prices which even poor people can generally afford to pay, they are the major suppliers of calories in the world. In most developing nations carbohydrates furnish over 75% of the total caloric intake. In the U.S. approximately 47% of the total calories consumed are in the form of carbohydrates (The Food and Nutrition Board, N.R.C., National Academy of Sciences, 1968). The effects of chemical treatments on the carbohydrates in plants are therefore of vital concern.

Vitamins

Vitamins are essential dietary constituents which are indispensable for optimum nutrition and health, and for normal growth and development. They promote growth, regulate body processes, and protect against disease. Even suboptimum quantities may help assure growth and give some protection against certain diseases connected with inadequate nutrition. But in quantities several times the amounts required for minimal protection, they may contribute to a state of nutrition that further safeguards health and enriches living. The Food and Nutrition Board of the National Research Council (1968) has recommended that a specific amount of the 13 known vitamins be included in the diet of the adult each day. These include the fat-soluble vitamins — A, D, E, and K — and the water-soluble ones — ascorbic acid,

biotin, folacin, niacin, pantothenic acid, riboflavin, thiamine, B_6, and B_{12}.

Vitamin A is essential to good nutrition and vigorous health in the adult as well as in the child. It stimulates growth and normal development. Lack of this vitamin is reflected in diseases of the eye, defective enamel in teeth, low resistance to infection, retarded growth, and loss of reproductive power.

β-Carotene, which will receive particular attention in this report, is known as the most important precursor of vitamin A and is the most common carotene found in green leaves, fruits, and roots. It is widely distributed in plants and occurs in animals such as marine invertebrates, marine fishes (in their ovaries), birds (in their eggs), insects (locusts and beetles), etc. All mammalian carotenes are of alimentary origin and there is no instance of any mammal manufacturing a specific carotene; but they have a general ability to convert certain carotenes, especially β-carotene, into vitamin A.

Minerals

Minerals are essential constituents of the hard and soft tissues and of the fluids of the body, and they assist in regulating body processes. They are widely distributed in all foods except the most highly refined, although the kinds and quantities may vary widely. The mineral elements needed by the body include calcium, chlorine, chromium, cobalt, copper, iodine, iron, magnesium, manganese, phosphorus, potassium, selenium, sodium, sulfur, zinc, and possibly fluorine, aluminum, nickel, vanadium, molybdenum, tin, and silicon. Many of these are referred to frequently as "trace elements" because only exceedingly small quantities are needed.

Of the macro minerals needed by the body, calcium, iron and iodine are most likely to be inadequate in the usual diet. Calcium is required for all tissues, but especially for bones and teeth. It is essential for coagulation of blood, rhythmic beat of the heart, irritability of nerves, cellular activity, and the maintenance of the acid-base balance of the body. Phosphorus is necessary for all tissues and like calcium, in addition to the structural role, is essential for energy metabolism. It helps maintain the neutrality of the blood, conduct of nerve impulses, and control of all activity. Iron is necessary to form hemoglobin, a substance in the blood which carries oxygen to the cells and plays an essential role in body functions and the activity of cells.

STIMULATION OF PLANT GROWTH AND YIELD BY CHEMICAL TREATMENTS

Relationships between pesticide usage and crop yields in various areas and countries are shown in Table 1. Pesticides are used extensively in countries whose sources of food are adequate. The countries that use the most ounces of pesticides per acre also have the highest production (Ennis et al., 1967). Soil applications of nematocides (Table 2) have increased yield production (National Academy of Sciences, 1968).

Flower cluster sprays of naphthaleneacetic acid (NAA), β-naphthoxyacetic acid (NOA), and 4-chlorophenoxyacetic acid (4-CPA) increased fruit setting and yields of greenhouse tomatoes (Zimmerman and Hitchcock, 1944; Murneek et al., 1944). Sprays and dusts of NAA, 4-CPA, and α-o-chlorophenoxypropionic acid (CPPA) increased pod set of snap beans (Wittwer and Murneek, 1946). The N-arylphthalamic acid treatments not only increased flower formation by tomato plants (Weaver, 1953; Teubner and Wittwer, 1955), but also increased their productivity under greenhouse conditions (Wittwer, 1960). The treatments induced large trusses of fruit for single cluster culture. Sprays of 200 ppm on young seedlings during the flower formation of the first cluster may increase the number of flowers and fruit severalfold.

2,3,5-Triiodobenzoic acid (TIBA) substantially increases the yield of soybeans under field conditions (Anderson, 1968; Colville, 1969). Increases of 2 to 5 bushels per acre have been reported and may go as high as 70%. Treatment increased pod set and accelerated maturity by 5 days. The most striking yield increases with TIBA on soybeans are achieved with plants grown at high fertility levels, narrow row spacing, and high plant populations. TIBA treatment also increased yields of Southern peas (Hipp and Cowley, 1969).

Gibberellic acid has been used to increase seed production of lettuce (Harrington, 1960), stalk length and yields of celery (Wittwer and Bukovac, 1957), yields of forcing rhubarb (Tompkins, 1965, 1966), and berry size of grapes (Itakura et al., 1956). A large-scale use for gibberellic acid (GA) may arise in sugar production from sugar cane (Tanimoto and Nickell, 1967). Maximum sucrose from sugar cane requires a high tonnage of stalks and a high degree of sucrose storage. High concentrations of GA stimulate stalk tissue production and increase sugar yields. It has been confirmed that GA promotes growth elongation of sugar-cane stalks with no appreciable change in sugar content. The net result is an increase in sugar yield per acre (Alexander, 1969). Yields of wheat are increased by soil treatment of 2-chloroethyltrimethyl-ammonium chloride (CCC) (Humphries, 1968; Steiger et al., 1969; and Wittwer, 1967). The same

TABLE 1

Areas and Nations Ranked in Order of Pesticide Usage Per Acre and in Order of Yields of Major Crops

Area or nation	Pesticide use		Yields	
	Ounces/acre	Rank	Pounds/acre	Rank
Japan	154.0	1	4,890	1
Europe	26.7	2	3,060	2
United States	21.3	3	2,320	3
Latin America	3.1	4	1,760	4
Oceania	2.8	5	1,400	5
India	2.1	6	730	7
Africa	1.8	7	1,080	6

Reproduced from Ennis, W. B., Jr., Jensen, L. L., Ellis, I. T., and Newson, L. D., in *The World Food Problem. A Report of the President's Science Advisory Committee,* Vol. III, 130, 1967.

TABLE 2

Yield Increase Following Soil Application of Nematocides

Crop plant	Increase (%)
Lima bean	35
Cotton	91
Soybean	126
Sugarbeet	175
Tobacco	13
Tomato	73

Reproduced from National Academy of Science, Control of plant-parasitic nematodes, in *Principles of Plant and Animal Pest Control,* Vol. 4, 172, 1968.

type of treatments promotes fruit setting in grapes (Coombe, 1965; Claus, 1965; and Barritt, 1970) and root growth in beans and tomatoes (Tognoni et al., 1967). It is promising for cotton production (Shafer, 1967 and Cyanamid International, 1966). A spray of 1,000 to 2,000 ppm N,N-dimethyl-aminosuccinamic acid (Alar-85®) on apple trees 10 to 14 days after bloom promoted an increased flowering in subsequent years (Batjer et al., 1964). Treatments of Alar-85 also increase yields and number of tubers of potatoes (Bodlaender and Algra, 1966; Humphries and Dyson, 1967). It has been reported (Humphries and Pathiyagoda, 1969) that morphactins (2-chloro-9-hydroxy-fluorene-9-carboxylic acid is a typical representative) applied at 1 to 10 ppm as a foliar spray increased yield of potatoes. The sprays of wheat seedling with 250 to 500 ppm Mendok (sodium 2,3-dichloroiso-butyrate) stiffened the straw and increased wheat production (Mohan Ram and Rustagi, 1969).

Chapman and Allen (1948) observed that the organic pesticide DDT stimulated plant growth when applied in the field. Foliar spray of high concentrations of DDT when applied in the greenhouse injured squash and cucumber, but lower concentrations stimulated the growth of tomato, bean, carrot, pea, corn, squash, and cucumber plants. Phosphate insecticides were later reported to stimulate plant growth (Wolfenbarger, 1948).

With an application of 10 to 20 ppm of 2,4-D applied as the ester at 2 weeks after emergence, a 35% increase in yield of shelled green beans was obtained (Wedding et al., 1956). However, in later tests, applications of 5 to 20 ppm applied in different years and at different locations resulted

in lima and snap bean yields increasing up to 70% or decreasing by 75%. Wort (1959) obtained increased yields of green beans, sugar beets, maize, and potatoes with foliar applications of dusts and sprays of subtoxic levels of 2,4-D. Payne et al. (1952) and Wort (1962) reported an increase in yield and quality of potatoes due to treatment with 2,4-D. Miller et al. (1962) found an increased growth of field beans with 0.5 to 1.0 ppm of 2,4-D. Huffaker et al. (1967) showed that the yield of wheat was increased with a dust treatment of 2,4-D mixed with micronutrients.

As to other herbicides, Bartley (1957) observed that corn was greener and taller when 2-chloro-4,6-bis(ethylamino)-s-triazine (Simazine®) was applied at rates up to 18 Kg/ha. Minshall (1960) showed that a subtoxic application of monuron increased the dry weight of bean plants grown in a controlled environment. Lorenzoni (1962) reported an increased growth rate in several cereals, with very low concentrations of Simazine. Peach and apple trees grew more when weeds were controlled with Simazine than with black plastic or hoeing (Ries et al., 1963). Freney (1965) established that Simazine increased corn growth under greenhouse conditions. At 0.06 ppm in solution culture, it increased the yield of corn tops by 36%. Citrus trees treated with either Simazine or 2-chloro-4-ethylamino-6-isopropylamino-s-triazine (Atrazine®) were also found to grow better than weeded controls (Goren and Monselise, 1966). Ries et al. (1968), in field studies in Michigan and Costa Rica, showed increases in yields of ryegrass, dry beans, and peas when treated with Simazine. They further showed that both Simazine and terbacil increased wheat yields in Michigan, but not in Mexico (Ries et al., 1970). However, McNeal et al. (1969) and Tweedy et al. (1971) did not report any beneficial effect of Simazine on yield of wheat. In greenhouse studies, Simazine application increased the growth of grains such as barley, winter wheat (Bastin et al., 1970), and oats (Steenbjerg et al., 1972), and in field studies with winter wheat (Swietochowski and Glabiszewski, 1967). Swietochowski and Glabisjewski (1967) also reported that sprays of 250 to 500 g Simazine/ha increased the depth of root growth of wheat. Wu et al. (1972a), using peas and sweet corn, found that foliar application of Simazine, 2-methoxy-4,6-bis(isopropylamino)-s-triazine (Prometone®), and 2-methylthio-4-ethylamino-s-triazine (Igran®) at 2 ppm and

2-chloro-4,6-bis(isopropylamino)-s-triazine (Propazine[®]) at 2 or 5 ppm significantly increased the fresh weight, but did not alter the total dry weight of the shoots of pea (Table 3). Although there was no significant effect of Simazine, Prometone, and Igran at a level of 5 ppm on fresh weight, the total dry weight of the shoots of pea was decreased (Table 3). Foliar applications of either 5 or 2 ppm of Simazine or Propazine increased the fresh weight of the shoots of sweet corn seedlings 5 days after application. In most cases, applications of s-triazines did not affect the fresh or dry weight of the shoots 15 days after treatment (Table 4).

Singh and Salunkhe (1972a) reported that Atrazine at 0.5 ppm caused a significant increase in height of bean plants on all the 3 days of measurement (Table 5). Plants treated with Atrazine at 0.1 ppm, terbutryne at 0.1 ppm or 0.5 ppm, and 2-methoxy-4-isopropylamino-6-butylamino-s-triazine (GS-14254) at 0.5 ppm showed significant differences in height, but only 10 or 20 days after application. Treated plants were taller than untreated plants 10 days after

application at both concentrations of Simazine. Atrazine at 0.5 ppm and terbutryne at 0.1 ppm significantly increased the fresh weight of leaf samples taken 5, 10, and 20 days after treatment (Table 5). A significant increase was also noted in the fresh weight of leaf samples taken 5 and 10 days after treatment with 0.1 and 0.5 ppm of Simazine. Compared with the control plants, the dry weights of leaves of the plants treated with 0.5 ppm Atrazine were significantly higher on all the 3 days of observations. Although there was no significant effect of terbutryne at 0.1 ppm on the dry weight of the leaves taken on the 5th day, the weights were increased significantly on the 10th and 20th days. Low concentrations of 2-methyl-thio-4,6-bis(isopropylamino)-s-triazine (Prometryne) stimulated growth of peas (Schulke, 1971).

Field sprays of 2-methyl-4,6-dinitrophenol (DNOC) on young winter rye increased the yield by 10% (Bruinsma, 1962). In a greenhouse study Bruinsma (1966) found that both the roots and shoots of DNOC-treated plants were smaller than

TABLE 3

Effects of Foliar Treatment of s-Triazines on Fresh and Dry Weight of Shoots of 3-week-old Pea Seedlings[†] (Under Growth Chamber Conditions)

		Days after treatment					
Treatment		5		10		15	
Compound	Concn. (ppm)	Fresh wt. g	Dry wt. g	Fresh wt. g	Dry wt. g	Fresh wt. g	Dry wt. g
Control		13.15	1.27	20.46	2.04	29.37	3.02
Simazine[®]	5	13.00ns	1.19*	19.97ns	1.89*	31.96*	3.13ns
	2	13.98*	1.31ns	22.85*	2.21*	32.51*	3.25*
Propazine[®]	5	14.02*	1.29ns	22.37*	2.12ns	31.85*	3.05ns
	2	13.94*	1.29ns	23.14*	2.19*	31.94*	3.13ns
Prometone[®]	5	12.90ns	1.19*	20.17ns	1.85*	28.95ns	2.86*
	2	14.55*	1.31ns	21.81*	2.07ns	31.65*	3.03ns
Igran[®]	5	13.01ns	1.17*	20.06ns	1.86*	30.44ns	2.83*
	2	14.09*	1.31ns	22.13*	2.08ns	31.69*	3.07ns

*Significantly different from control at 0.05 level.
ns Not significantly different from control at 0.05 level according to Tukey's ω-procedure.
[†] Average value per pot for six plants.

Reproduced from Wu, M. T., Singh, B., and Salunkhe, D. K., *J. Exp. Bot.*, 23, 793, 1972a.

TABLE 4

Effects of Foliar Treatment of *s*-Triazines on Fresh and Dry Weight of Shoots of Sweet Corn Seedlings[†] (Under Greenhouse Conditions)

| Treatment | | Days after treatment | | | | | |
| | | 5 | | 10 | | 15 | |
Compound	Concn. (ppm)	Fresh wt. g	Dry wt. g	Fresh wt. g	Dry wt. g	Fresh wt. g	Dry wt. g
Control		11.64	1.52	16.38	2.21	21.87	2.84
Simazine	5	13.48*	1.52ns	18.64*	2.20ns	22.65ns	2.87ns
	2	13.57*	1.57ns	18.49*	2.14ns	23.70*	2.96ns
Propazine	5	13.24*	1.57ns	18.05*	2.18ns	22.72ns	2.93ns
	2	13.31*	1.53ns	16.07ns	1.99*	22.20ns	2.90ns
Prometone	5	11.94ns	1.31*	18.43*	2.26ns	22.04ns	2.90ns
	2	12.20ns	1.34*	18.56*	2.22ns	21.95ns	2.85ns
Igran	5	12.89*	1.34*	18.04*	2.11ns	22.49ns	2.87ns
	2	13.64*	1.47ns	18.11*	2.10ns	22.74ns	2.93ns

*Significantly different from control at 0.05 level.
ns Not significantly different from control at 0.05 level.
[†] Average value per pot for six plants.

Reproduced from Wu, M. T., Singh, B., and Salunkhe, D. K., *J. Exp. Bot.*, 23, 793, 1972a.

controls 14 days after treatment, but appreciably larger than controls 56 days after treatment. Kesner and Ries (1968) demonstrated that very low concentrations of diphenamid enhanced the growth of tomato plants. Wiedman and Appleby (1972) in a greenhouse study showed that 11 out of 16 herbicides representing many different chemical families significantly increased oat seedling shoot and root growth. Deyton (1973) reported that use of several herbicides including bensulide, triflurolin dinoseb, and naptalam increased vegetable yields in the field. Bensulide increased the total yields of muskmelons by 48% and triflurolin increased bean yield by 26%. Wyse (1974) found that subtoxic levels of Atrazine applied prior to planting increased total seed yield of field beans.

As to other chemicals, growth of sugar cane was increased by the application of two substituted hydrocarbon insecticides (Jaiswal et al., 1973). Fungicides, organic solvents, and long-chain alcohols have also been reported to stimulate plant growth (Southam and Ehrlich, 1943; Nutman and Roberts, 1962; and Morre et al., 1965). Recently, growth substances derived from petroleum have been repeatedly shown to increase plant growth (Gvozdenko, 1966).

INCREASE IN PROTEIN CONTENT OF PLANTS BY CHEMICAL TREATMENTS

2,4-D

Erickson et al. (1948) showed that spraying wheat with 2,4-D (2,4-dichlorophenoxy acetic acid) changed the chemical composition of the grain. The protein content in 7 wheat cultivars increased in direct relation to the amount of 2,4-D applied. Payne et al (1953) found that 2,4-D treatment increased protein 27% in potato tubers. An application of 560 g/ha of 2,4-D to potato foliage resulted in more protein in potatoes both at harvest and after 60 days of storage (Yasuda et al., 1956). Spraying of 100 ppm 2,4-D on buckwheat increased soluble and total nitrogen in both stems and roots, whereas the nitrogen content of both the stems and roots remained unchanged (Wort, 1959). Shushu (1966) measured increased

TABLE 5

Effects of Foliar Treatment of Atrazine, Simazine, Terbutryn, and GS-14254 on Fresh Weight and Dry Weight of the Leaves and Height of Bush Bean Plants 5, 10, and 20 days After Treatment (Under Growth Chamber Conditions)

Days after treatment	Chemical	Concn. (ppm)	Plant height (cm)	Fresh wt. of the leaves/ plants (g)	Total dry wt. of the leaves/ plants (g)
5	Control		12.97	12.53	1.36
	Atrazine®	0.1	12.47ns	12.62ns	1.43ns
		0.5	13.15*	15.97*	1.54*
	Simazine®	0.1	12.95ns	14.16*	1.18ns
		0.5	12.74ns	14.78*	1.29ns
	Terbutryn	0.1	12.87ns	15.69**	1.30ns
		0.5	13.12ns	12.97ns	1.31ns
	GS-14254	0.1	12.54ns	13.21ns	1.06*
		0.5	12.73ns	12.97ns	1.22ns
10	Control		15.79	23.62	2.75
	Atrazine	0.1	18.24*	23.68ns	2.89ns
		0.5	19.86**	25.46*	3.55*
	Simazine	0.1	17.65*	27.99*	2.79ns
		0.5	17.91*	28.34*	2.78ns
	Terbutryn	0.1	17.87*	26.86*	3.25*
		0.5	16.33ns	23.52ns	2.66ns
	GS-14254	0.1	16.54ns	22.75ns	3.14ns
		0.5	19.13**	24.01ns	2.67ns
20	Control		29.99	47.01	5.78
	Atrazine	0.1	34.11*	54.03*	7.07*
		0.5	34.51**	54.09*	7.22**
	Simazine	0.1	29.91ns	49.97ns	5.14ns
		0.5	32.01ns	47.04ns	5.76ns
	Terbutryn	0.1	33.95ns	55.56*	7.26*
		0.5	34.29*	46.36ns	5.99ns
	GS-14254	0.1	33.52ns	51.57ns	6.88*
		0.5	34.62*	58.96**	6.26ns

*Significantly different at the 0.05 level.
ns Not significantly different from control at 0.05 level.
**Significantly different from control at 0.01 level.

Reproduced from Singh, B., Salunkhe, D. K., and Lipton, S. H., *J. Hortic. Sci.,* 47, 441, 1972a.

nitrogen in grains treated with phenoxy acids without reductions in yield. Huffaker et al. (1967) found that treatment with 13 g/ha 2,4-D mixed with micronutrients increased the protein content of wheat and barley without decreasing yield. Khripunova (1967) observed an increased protein content of spring wheat accompanied by increased yield with applications of 0.5 to 1.3 kg/ha of 2,4-D made to the soil prior to sowing the wheat.

Sell et al. (1949) reported that stems of red kidney plants treated with 2,4-D contain approximately twice as much protein as stems of untreated plants. The amino acids, calculated as the percent of crude protein, showed the greatest differences in the quantity of amino acids of the stem tissue to occur in aspartic acid, lysine, valine, methionine, and phenylalanine. The variations in the contents of amino acids indicate that the character of the protein is different in the treated stems than in the controls.

Payne et al. (1952) found that 2,4-D treatment decreased the quantity of 11 amino acids in potato tubers, with no increase in free glutamic acid. Gruzdev et al. (1969) observed increased nitrogen and percentage of glutamic acid and proline in grains without reduction in yield when treated with 2,4-D.

s-Triazines
Soil Treatments on Protein Content

Bartley (1957) first observed that corn treated with Simazine as an herbicide appeared to be greener in color. Working with herbicidal levels, Gast and Grob (1960) detected more crude protein in forage corn treated with 1 to 6 kg/ha of Atrazine or Simazine. Ries et al. (1963) found that peach trees treated with the herbicide mixture of Simazine and 3-amino-1,2,4-triazole plus ammonium thiocyanate (amitrole-T) had a higher leaf nitrogen content and more growth than trees where the weeds were controlled by hand hoeing or a black plastic mulch. Apple trees had higher leaf nitrogen when treated with Simazine and amitrole-T than with no weed control. Both peach and apple trees had higher leaf nitrogen after herbicide treatments than after supplemental nitrogen treatments, indicating herbicidal influence on the nitrogen metabolism of the trees. DeVries (1963) demonstrated increased nitrogen uptake by corn in Simazine-treated soils, and corn plants showing nitrogen deficiency symptoms

overcame these symptoms when Simazine was applied at 15 lb/A. Ries and Gast (1965) reported that the addition of Simazine to nutrient solutions of corn increased the percent of N and total mg of N in one test regardless of the N level in the solution. Under the low N level, the magnitude of this increase was 90% with 0.5 ppm Simazine. However, in a second experiment, under environmental conditions more favorable for corn growth, the total mg of N was not increased, although the percent nitrogen in the shoots was increased at the lower nitrogen level. Corn plants treated with hydroxysimazine responded similarly to the controls.

Freney (1965) demonstrated that Simazine applications of 1.5 ppm to soil in pots in the greenhouse increased dry matter yield and uptake of nitrogen in corn only when additional nitrogen was applied to the soil. The proportions of protein and nonprotein nitrogen in the topgrowth were not affected. Application of Simazine at this rate without additional nitrogen resulted in increased leaf area, but did not significantly affect yield. Freney pointed out that Simazine provided a stimulus to the plant early in its life and enabled it to grow faster than the control plant. When the supply of available nitrogen was exhausted, the Simazine effect decreased. When incubated with soil, Simazine did not increase mineralization of soil organic nitrogen, nor did it have any effect on immobilization of mineral nitrogen. It had no effect on the yields of roots. Freney suggested that Simazine increased plant growth by a direct effect on plant metabolism and not through any interaction with the soil. Eastin and Davis (1967) showed that in field, soil culture, and nutrient culture experiments Atrazine increased percent total nitrogen in all species studied, whether resistant, intermediate, or susceptible to 1 lb/A of Atrazine. However, the nitrogen content per plant was either unaffected or decreased by the Atrazine treatments.

In most species studied 1 lb/A of Atrazine increased the percent of protein and nitrate nitrogen. Atrazine susceptibility apparently was related to its effect on percent nonprotein and ammonia in the species tested. In resistant species 1 lb/A of Atrazine had no effect on the percent of nonprotein nitrogen and decreased the percent of ammonia nitrogen, but increased both fractions in intermediate and susceptible species. Hydroxyatrazine, a degradation product of Atrazine, at the same rate had no effect on nitrogen content of corn.

Gramlich and Davis (1967) found that corn and Johnson grass plants treated with high rates of Atrazine always contained less nitrogen (mg/plant) than untreated plants. Atrazine-treated corn plants contained higher percentages of both 80% ethanol-soluble and insoluble nitrogen than did the check plants. Percentage increases in both fractions were proportional to the rate of Atrazine treatment. Nitrate percentage was also increased, but free ammonium content was not significantly affected. Fink and Fletchall (1967) observed that Atrazine and Simazine caused an increase in the percent of nitrate in forage approximately 5 weeks after treatment. Percent of nitrogen in corn forage increased sharply approximately 5 weeks after planting due to herbicidal treatments, especially on plots not treated with nitrogen fertilizer. Ries et al. (1967) demonstrated that Simazine caused an accumulation of protein in rye plants and the level of nitrate reductase activity increased when plants were grown in a medium containing nitrate, whereas no effects were noticed if grown in a medium containing ammonium ions.

Field applications of Simazine under the vastly different environmental conditions of Michigan and Costa Rica caused similar responses. Both the yield and protein content were increased in ryegrass forage and in pea and bean seed, and the protein content was increased in bean seed, rice foliage, alfalfa forage, and oats (Ries et al., 1968). However, in an extensive study McNeal et al. (1969) showed no increase in the protein content of spring wheat without a decrease in yield. Soil treatments of s-triazines increased the protein contents of potatoes, oats, and radishes (Mazur and Kasecka, 1969). Wheat, barley, clover, and alfalfa accumulated more crude protein after sprays of 0.17% Simazine (Sojkowski et al., 1969). Allinson and Peters (1970) noted that sublethal applications of Simazine significantly increased the crude protein contents of a number of forage crops.

Extensive tests with wheat showed appreciable increases in protein yields per hectare in Michigan with both Simazine and terbacil application. However, any increase in protein content of wheat grain grown under the irrigated Mexican conditions was accompanied by a decrease in yield (Ries et al., 1970). Increases in protein content of wheat and barley were noted by Bastin et al. (1970) with

sprays of 0.1 to 0.02 ppm Simazine. Miller and Mikkelson (1970) reported that an application of 560 g/ha of Simazine increased the protein yield of rice more than 33 kg/ha of nitrogen. Increased protein in rice grain has been shown with granular applications and sprays applied at panicle initiation or at heading (Dedatta et al.,1972). The increases were 13, 17, and 18% more crude protein for Simazine, Simetryne®, and benzomarc, respectively.

Vergara et al. (1970) reported that while the treatment of Simazine to flooded soil at flowering time increased the protein content of rice, it was accompanied by a decrease in grain yield, which can be attributed to increased sterility. This decrease in grain yield consequently lowers the total grain protein production per crop. They also found that application of Simazine to young rice plants grown in culture solution increased the nitrogen content of the blades, sheaths, culms, and roots. The increase, however, is not the result of an increase in nitrogen uptake by the rice plant. The reduction in dry weight accounts for the increase in the percent of nitrogen. Tweedy et al. (1971), working with wheat and sorghum, observed that field application of Simazine has no effect on protein content. However, in one year a 35% increase in protein was attributed to subtoxic applications when nitrogen was also applied at 75 kg/ha. With application of fertilizer containing Atrazine or a mixture of Atrazine and Prometryne, corn plants absorbed a greater amount of nitrogen (Sosnovaya, 1971).

Foliar Treatments on Protein Content

In a study of peas and sweet corn Wu et al. (1972a) noticed that all the s-triazine compounds at concentrations of either 5 or 2 ppm increased the total N and soluble protein contents, and decreased the starch and sugar contents of pea seedlings 5 days after treatment (Table 6). Soluble protein contents in the leaves of treated seedlings were higher than those of the controls on the 10th and 15th days; however, the intensity of the effects seems to be decreased at that time. s-Triazine compounds at either 5 or 2 ppm increased the total nitrogen and soluble protein, and decreased the starch and soluble sugar contents of sweet corn seedlings. Although the influence of some treatments such as 5 or 2 ppm Simazine, Prometone, and Igran was still apparent on the 15th day, the intensity of the influence

TABLE 6

Effects of Foliar Treatment of *s*-Triazines on Chemical Composition of the Leaves of 3-week-old Pea Seedlings[†] (Under Greenhouse Conditions)

| Treatment | | Days after treatment | | | | | | | | | | | |
| | | 5 | | | | 10 | | | | 15 | | | |
Compound	Conc. (ppm)	Total N %	Soluble protein %	Starch %	Soluble sugars %	Total N %	Soluble protein %	Starch %	Soluble sugars %	Total N %	Soluble protein %	Starch %	Soluble sugars %
Control		4.8	21.6	5.5	9.8	4.0	20.1	5.9	9.2	3.6	18.8	5.7	9.5
Simazine	5	5.2**	26.7**	3.6**	5.2**	4.5**	22.9**	4.0**	6.0**	4.1**	20.6**	4.4**	7.5**
	2	5.2**	26.3*	3.9**	5.7**	4.5**	22.2**	4.1**	6.4**	4.2**	21.3**	4.2**	6.9**
Propazine	5	5.4**	25.9**	3.2**	5.8**	4.7**	21.9*	3.5**	6.6**	4.3**	20.4*	3.9**	7.9*
	2	5.5**	26.3*	3.5**	5.8**	4.6**	22.0*	3.9**	6.4**	4.3**	21.0**	4.0**	7.3**
Prometone	5	6.0**	26.9**	4.1**	4.3**	4.9**	23.1**	4.5**	5.9**	4.4**	21.1**	4.8*	7.6*
	2	5.9**	25.4*	3.6**	5.0**	5.0**	22.0*	4.0**	6.1**	4.5**	20.9**	4.7*	7.4**
Igran	5	5.8**	26.3*	4.3**	5.6**	4.8**	23.2**	4.3**	6.5**	4.3**	20.4*	4.6*	7.4**
	2	5.6**	25.4*	4.1**	5.7**	4.5**	22.6**	4.4**	6.3**	4.0*	20.7*	4.6*	7.4**

*Significantly different from control at 0.05 level.
**Significantly different from control at 0.01 level.
†Percent of dry weight of the leaflets.

Reproduced from Wu, M. T., Singh, B., and Salunkhe, D. K., *J. Exp. Bot.*, 23, 793, 1972a.

decreased (Table 7). Singh et al. (1972a) found that leaf samples from s-triazine-treated bush bean plants did not differ significantly from the untreated controls (Table 8). The application of each of the four herbicides resulted in decreased contents of starch, total and reducing sugars, and increased contents of total and reducing sugars and of total protein and soluble amino acids in the leaves of treated plants (Table 8).

By using foliar treatments, Singh et al. (1972c) found that Simazine or Terbutryn at the rates of 5 ppm and Prometone at 2 ppm caused significant increases in the amounts of total nitrogen and soluble protein in pea seeds when applied to the leaves of 45-day-old plants (Table 9). The starch and total sugar contents were decreased. Foliar treatments of 5 ppm of Simazine, Terbutryn, and Prometone to 60-day-old sweet corn plants significantly increased total N and soluble protein, but decreased starch and total sugars. The total N and soluble amino acids were significantly increased by the application of 0.5 ppm of Atrazine, Simazine, and Terbutryn to 35 to 42-day-old bean plants under both field and growth room conditions. These same treatments did not have any significant effect on nitrate-N total sugars or starch (Table 10). The total fat content in the seeds of peas and sweet corn was not significantly affected by s-triazine compounds (Table 9). Study on spinach under growth room conditions indicated that sublethal concentrations of s-triazine induced production of 12 to 13% more proteins in the leaves (Table 11). Starch content of the leaves was not affected; however, total sugars were lower (11 to 17%) in treated than in untreated plants.

Protein Quality: Amounts and Patterns of Amino Acids

After 2 years of experiments, Salunkhe et al. (1971a) reported that the treatment of plants with Simazine, Atrazine, Igran, or 2-methylthio-4-ethyl-amino-6-isopropylamino-s-triazine (Ametryne®) at the rate of 2 lb/A decreased the total nitrogen and soluble protein content in the seeds of peas (Table 12). Simazine or Igran at the rates of 0.5 and 0.125 lb/A, Propazine at the rates of 2 and 0.5 lb/A, Prometone at all three rates, and Ametryne at 0.125 lb/A significantly increased the total N and soluble protein contents while decreasing the starch and soluble sugar contents of pea seeds. Atrazine at the rates of 0.5 and 0.125 lb/A,

Propazine at the rate of 0.125 lb/A, GS-14254 and Prometryne at all three rates, and Ametryne at 0.5 lb/A had no effect on the protein and total N of the peas in the first year experiment. In the second year experiments only Simazine, Propazine, Prometone, Igran, and Ametryne were used. The data in Table 13 indicate that these chemicals had more or less similar effects on the chemical compositions of pea seeds. Three concentrations of each of the s-triazines (Simazine, Atrazine, Propazine, GS-14254, Prometone, Prometryne, Igran, and Ametryne) were applied in the first year for sweet corn. Simazine at the rates of 4 and 1 lb/A and Prometone, Igran, and Ametryne at the rate of 4 lb/A significantly increased the total N and soluble protein and decreased the starch contents of sweet corn seeds (Table 14). Further support is given by the results of the second year experiments (Table 15).

Although statistical analysis was not possible in both years, the responses of pea and sweet corn plants were similar with respect to amino acid patterns in the seeds (Tables 16 and 17). In every instance total amino acids of the treated plants were more than those of the untreated controls; and in most cases individual amino acids were also higher in amount. In almost every treatment, isoleucine, histidine, and cystine were lower in amount than the controls in pea seeds in both the first and second years. In sweet corn seeds, however, only glutamic acid was lower in both years.

The effects of Simazine and Prometone on amino acid patterns of peas and sweet corn seeds are presented in Table 18 (Singh et al.,1972c). In every case total amino acid content was higher in the Simazine- or Prometone-treated plants than the untreated controls. Compared to the controls, Simazine and Prometone increased the concentrations of most of the amino acids in pea seeds, except that histidine and cystine in Simazine-treated plants and isoleucine, methionine, and histidine in Prometone-treated plants were lower. In sweet corn the amounts of each individual amino acid were higher, except for hydroxyproline and glutamic acid in Simazine-treated plants and valine, methionine, and hydroxyproline in Prometone-treated plants.

Hiranpradit et al. (1972a) recently showed that Atrazine at 1.5 ppm concentration did not significantly affect plant dry weight, but significantly

TABLE 7

Effects of Foliar Treatment of s-Triazines on Chemical Composition of the Leaves of 3-week-old Sweet Corn Seedlings (Under Greenhouse Conditions)

| Treatment | | Days after treatment | | | | | | | | | | | |
| | | 5 | | | | 10 | | | | 15 | | | |
Compound	Conc. (ppm)	Total N %	Soluble protein %	Starch %	Soluble sugars %	Total N %	Soluble protein %	Starch %	Soluble sugars %	Total N %	Soluble protein %	Starch %	Soluble sugars %
Control		1.2	5.2	18.0	5.9	1.5	6.9	17.6	6.4	2.0	8.9	18.2	7.6
Simazine	5	1.6**	7.1**	14.1**	5.0**	1.9**	8.2**	15.2**	5.8*	2.5**	9.9**	16.8*	7.3ns
	2	1.5**	6.8**	14.7**	5.2**	1.9**	7.7**	16.0*	5.4**	2.3**	9.6*	16.5**	7.5ns
Propazine	5	1.5**	7.0**	13.9**	4.8**	1.8**	8.3**	15.4**	5.3**	2.3**	9.7*	16.6**	7.6ns
	2	1.4**	6.5**	14.4**	5.0**	1.8**	7.6**	15.6*	5.6*	2.3**	9.7*	16.1**	7.4ns
Prometone	5	1.8**	7.4**	14.2**	4.8**	2.1**	8.9**	15.6**	5.5**	2.5**	10.0**	17.0*	7.5ns
	2	1.6**	6.6**	14.5**	4.9**	2.0**	7.8**	15.5**	5.3**	2.4**	9.7*	16.4**	7.4ns
Igran	5	1.9**	7.3**	13.6**	4.7**	2.1**	8.2**	15.5**	6.0*	2.5**	9.9**	16.7*	7.4ns
	2	1.7**	7.0**	14.3**	4.9**	2.0**	8.0**	15.8*	5.6*	2.3*	9.7*	16.4**	7.7ns

*Significantly different from control at 0.05 level.
**Significantly different from control at 0.01 level.
ns Not significantly different from control at 0.05 level.
† Percent of dry weight of the leaflets.

Reproduced from Wu, M. T., Singh, B., and Salunkhe, D. K., *J. Exp. Bot.*, 23, 793, 1972a.

TABLE 8

Effects of Foliar Treatment of Atrazine, Simazine, Terbutryn, and GS-14254 on Chemical Composition of Leaves of Bush Bean Plants 5, 10, and 20 days After Treatment (Under Growth Chamber Conditions)

Days after treatment	Chemical	Concn. (ppm)	Starch (mg/g dry wt.)	Total sugar (mg/g dry wt.)	Reducing sugar (mg/g dry wt.)	Total protein (mg/g fresh wt.)	Amino acid (mg/g dry wt.)	Nitrate (mg/g dry wt.)
5	Control		27.7	7.9	5.7	16.1	3.2	24
	Atrazine	0.1	21.0**	5.8**	4.6*	20.5*	4.5*	26ns
		0.5	25.4ns	4.9**	5.0ns	39.2**	4.1**	25ns
	Simazine	0.1	22.9*	6.5**	3.6**	19.5*	5.8*	28ns
		0.5	20.0**	5.8**	4.4*	19.4*	5.9**	29ns
	Terbutryn	0.1	23.2*	6.7*	5.1ns	40.0**	4.9**	28ns
		0.5	18.5**	5.8**	3.1**	26.3**	3.5*	30ns
	GS-14254	0.1	19.8**	7.9ns	5.5ns	33.5**	3.9*	30ns
		0.5	18.5**	4.9**	4.1*	28.3**	4.4*	26ns
10	Control		36.6	4.5	2.4	12.3	3.3	37
	Atrazine	0.1	25.0**	4.4ns	1.9*	22.3**	5.3**	36ns
		0.5	28.0**	3.0**	1.8**	18.7**	5.5**	40ns
	Simazine	0.1	20.6**	3.7**	2.5ns	19.3**	4.3**	35ns
		0.5	21.2**	3.4*	1.7**	15.7*	5.8**	39ns
	Terbutryn	0.1	26.6**	3.5**	1.5*	22.1**	4.1*	41ns
		0.5	27.4**	4.1ns	1.9ns	21.4**	4.3**	43ns
	GS-14254	0.1	25.4**	4.1ns	1.8*	21.5**	5.0**	36ns
		0.5	26.9**	3.5**	2.0ns	20.7**	5.8**	39ns
20	Control		39.2	6.4	3.0	19.4	4.3	29
	Atrazine	0.1	26.4**	5.5ns	2.1*	27.7**	5.5**	26ns
		0.5	24.7**	5.0*	2.2*	30.6**	5.0*	28ns
	Simazine	0.1	28.3**	5.8ns	2.0*	26.8**	5.6**	33ns
		0.5	22.5**	4.3**	1.9*	24.4**	5.8**	30ns
	Terbutryn	0.1	19.9**	5.2*	1.8**	23.7**	5.5**	33ns
		0.5	22.2**	5.7ns	1.6**	22.8*	5.7**	31ns
	GS-14254	0.1	31.6**	5.0*	2.1*	21.4*	6.2**	32ns
		0.5	25.5**	4.9*	2.4ns	23.4**	6.3**	28ns

*Significantly different from control at 0.05 level.
**Significantly different from control at 0.01 level.
ns Not significantly different from control at 0.05 level.

Reproduced from Singh, B., Salunkhe, D. K., and Lipton, S. H., *J. Hortic. Sci.,* 47, 441, 1972a.

TABLE 9

Effects of Foliar Treatment of s-Triazines on Dry Weight, Protein, Total N, Starch, and Sugars and Fat in the Seeds of Peas and Sweet Corn 30 days After Application[a] (Under Field Conditions)

Crop	Treatment	Concn. (ppm)	Dry wt. %	Total N %	Soluble protein %	Starch %	Total sugars %	Crude fat %
Peas	Control	0	23.3	4.4	12.9	48.4	6.6	2.7
	Simazine	5	22.9	5.2*	15.6*	44.2*	5.7*	2.9
	Terbutryn	5	22.4	5.1*	14.9*	45.2*	5.7*	
	Prometone	2	22.7	5.1*	14.6*	45.0*	5.5*	2.9
Sweet corn	Control	0	28.3	1.2	4.6	59.9	14.3	5.9
	Simazine	5	28.5	1.4*	5.3*	49.0*	12.4*	5.9
	Terbutryn	5	27.9	1.3*	5.2*	49.3*	12.6*	
	Prometone	5	28.4	1.4*	5.3*	49.5*	12.2*	5.9

[a]All data expressed on dry weight basis.
*Significantly different from control at 0.05 level.

Reproduced from Singh, B., Vadhava, O. P., Wu, M. T., and Salunkhe, D. K., *J. Agric. Food Chem.*, 20, 1256, 1972c.

TABLE 10

Effects of Foliar Treatment of s-Triazines on Total N, Soluble Amino Acids, Starch, and Sugars in the Pods of Bush Beans[a] (Under Field and Growth Chamber Conditions)

Treatment	Concn. (ppm)	Dry wt. %	Total N %	Nitrate μg/g	Soluble amino acids mg/100 g	Starch %	Total sugars %
				Field experiments[b]			
Control	0	8.8	1.8	38	33.3	69.3	5.9
Atrazine	0.5	9.5	2.4*	40	38.7*	68.5	5.9
Simazine	0.5	9.0	2.0*	36	39.0*	68.9	5.2
Terbutryn	0.5	8.5	2.2*	37	39.2*	69.1	5.9
				Growth chamber experiments[c]			
Control	0	9.1	2.8	32	25.0	63.3	4.5
Atrazine	0.5	8.8	3.2*	29	39.1*	62.8	4.6
Simazine	0.5	9.5	3.5*	30	36.2*	67.6	4.7
Terbutryn	0.5	9.0	3.7*	32	31.2*	64.2	4.3

[a]All data expressed on dry weight basis.
[b]Bean pods were harvested 28 days after application.
[c]Bean pods were harvested 18 days after application.
*Significantly different from control at 0.05 level.

Reproduced from Singh, B., Vadhava, O. P., Wu, M. T., and Salunkhe, D. K., *J. Agric. Food Chem.*, 20, 1256, 1972c.

TABLE 11

Effects of Foliar Treatment of *s*-Triazines on Total N, Soluble Amino Acids, and Sugars on the Leaves of Spinach 18 days After Application[a] (Under Growth Chamber Conditions)

Treatment	Concn. (ppm)	Dry wt. %	Total N %	Soluble protein[b] mg/g	Starch %	Nitrate μg/g	Soluble amino acids mg/g	Total sugars %
Control	0	9.4	4.1	12.9	28.1	31	4.3	7.6
Atrazine	0.5	9.7	5.0*	14.1*	27.9	29	5.1*	6.8*
Simazine	0.5	9.2	5.2*	13.7*	27.5	28	5.6*	6.3*
Terbutryn	0.5	9.2	5.5*	13.9*	28.5	32	4.9*	6.7*

[a]All data other than soluble protein are expressed on dry weight basis.
[b]Expressed on fresh weight basis.
*Significantly different from control at 0.05 level.

Reproduced from Singh, B., Vadhava, O. P., Wu, M. T., and Salunkhe, D. K., *J. Agric. Food Chem.,* 20, 1256, 1972c.

TABLE 12

Effects of Soil Treatment of *s*-Triazine Compounds on the Dry Weight and Chemical Composition of Peas (1st year) (Under Field Conditions)

Compound	Concn. (lb/A)	Dry wt. %	Total N %	Soluble protein %	Starch %	Sugar %
Control		35.04[a]	4.31	14.20	51.33	5.54
Simazine	2	35.11ns	4.06*	12.91*	49.06*	5.39
	0.5	35.03ns	5.28*	17.02*	47.65*	4.61*
	0.125	35.17ns	4.94*	16.47*	48.10*	5.43*
Atrazine	2	34.95ns	4.00*	13.33*	48.71*	5.04*
	0.5	34.89ns	4.27ns	14.05ns	51.90ns	5.60ns
	0.125	35.14ns	4.36ns	14.52ns	50.33ns	5.71ns
Propazine	2	35.09ns	4.66*	15.89*	49.38*	5.21*
	0.5	35.04ns	4.70*	16.07*	49.05*	5.24*
	0.125	35.00ns	4.29ns	13.80ns	52.21ns	5.64ns
GS-14254	2	35.22ns	4.27ns	14.04ns	51.94ns	5.50ns
	0.5	35.17ns	4.38ns	14.66ns	50.61ns	5.39ns
	0.125	35.20ns	4.33ns	14.11ns	51.44ns	5.62ns
Prometone	2	35.06ns	4.62*	16.35*	49.03*	4.33*
	0.5	34.93ns	4.98*	16.82*	46.97*	4.75*
	0.125	34.99ns	4.87*	16.50*	47.73*	4.38*
Prometryne	2	35.14ns	4.22ns	14.07ns	52.35ns	5.47ns
	0.5	35.02ns	4.40ns	14.71ns	50.69ns	5.02ns
	0.125	35.15ns	4.36ns	14.43ns	51.18ns	5.60ns
Igran	2	35.28ns	4.13*	13.58*	52.37ns	5.48ns
	0.5	35.13ns	4.80*	16.72*	48.15*	4.47*
	0.125	34.88ns	4.71*	16.64*	48.76*	4.64*

TABLE 12 (continued)

Effects of Soil Treatment of s-Triazine Compounds on the Dry Weight and Chemical Composition of Peas (1st year) (Under Field Conditions)

Treatment

Compound	Concn. (lb/A)	Dry wt. %	Total N %	Soluble protein %	Starch %	Sugar %
Ametryne	2	35.06ns	3.94*	13.04*	51.62ns	5.01*
	0.5	35.21ns	4.40ns	14.71ns	48.21*	4.72*
	0.125	35.19ns	4.71*	15.94*	47.70*	5.07*

[a] Each value represents the mean of 3 samples. Triplicate determinations of each 3 samples were made. Dry weight expressed as percent of fresh weight of the seeds.
*Significantly different from control at 0.05 level.
ns Not significantly different from control at 0.05 level.

Reproduced from Salunkhe, D. K., Wu, M. T., and Singh, B., *J. Am. Soc. Hortic. Sci.*, 96, 489, 1971a.

TABLE 13

Effects of Soil Treatment of s-Triazines On the Dry Weight and the Chemical Composition of Peas (2nd year) (Under Field Conditions)

Treatment

Compound	Concn. (1b/A)	Dry wt. %	Total N %	Soluble protein %	Starch %	Sugar %
Control		36.93[a]	4.25	12.19	48.51	6.73
Simazine	0.5	37.11ns	5.09*	15.89*	44.11*	4.98*
	0.125	36.98ns	4.90*	15.03*	44.62*	5.21*
Propazine	2	36.70ns	4.77*	14.93*	44.94*	5.44*
	0.5	37.03ns	4.72*	14.64*	45.12*	5.29*
Prometone	2	36.94ns	4.70*	14.52*	44.87*	5.03*
	0.5	36.81ns	4.92*	14.98*	44.70*	5.49*
	0.125	37.15ns	4.84*	15.11*	43.97*	5.54*
Igran	0.5	36.90ns	4.83*	15.25*	44.26*	5.06*
	0.125	36.84ns	4.70*	14.65*	45.51*	5.24*
Ametryne	0.125	36.76ns	4.75*	15.20*	45.26*	5.61*

[a]Each value represents the mean of 3 samples. Triplicate determinations of each of 3 samples were made. Dry weight expressed as percent of fresh weight; other components, as percent of dry weight of the seeds.
*Significantly different from control at 0.05 level.
ns Not significantly different from control at 0.05 level.

Reproduced from Salunkhe, D. K., Wu, M. T., and Singh, B., *J. Am. Soc. Hortic. Sci.*, 96, 489, 1971a.

TABLE 14

Effects of Soil Treatment of *s*-Triazine Compounds on the Dry Weight and the Chemical Composition of Sweet Corn (1st year) (Under Field Conditions)

Treatment

Compound	Concn. (1b/A)	Dry wt. %	Total N %	Soluble protein %	Starch %	Sugar %
Control		44.52[a]	1.14	4.49	54.6	11.75
Simazine	4	44.31ns	1.25*	4.86*	50.7*	11.55ns
	1	44.40ns	1.22*	4.73*	51.0*	11.63ns
	0.25	44.41ns	1.13ns	4.51ns	53.9ns	11.76ns
Atrazine	4	44.37ns	1.20*	4.67*	52.0*	11.69ns
	1	44.56ns	1.15ns	4.43ns	54.8ns	11.70ns
	0.25	44.59ns	1.12ns	4.49ns	55.1ns	11.81ns
Propazine	4	44.47ns	1.17ns	4.53ns	53.7ns	11.77ns
	1	44.38ns	1.17ns	4.40ns	54.4ns	11.68ns
	0.25	44.51ns	1.13ns	4.49ns	54.9ns	11.73ns
GS-14254	4	44.40ns	1.15ns	4.51ns	53.8ns	11.70ns
	1	44.54ns	1.11ns	4.54ns	55.0ns	11.59ns
Prometone	4	44.52ns	1.24*	4.79*	51.4*	11.35*
	1	44.43ns	1.13ns	4.47ns	54.7ns	11.57ns
	0.25	44.60ns	1.13ns	4.50ns	55.3ns	11.74ns
Prometryne	4	44.53ns	1.11ns	4.48ns	54.1ns	11.80ns
	1	44.64ns	1.13ns	4.55ns	55.7ns	11.66ns
	0.25	44.57ns	1.17ns	4.53ns	54.0ns	11.71ns
Igran	4	44.30ns	1.21*	4.69*	53.4ns	11.68ns
	1	44.37ns	1.14ns	4.40ns	55.2ns	11.76ns
	0.25	44.63ns	1.16ns	4.47ns	54.8ns	11.70ns
Ametryne	4	44.50ns	1.21*	4.76*	51.6*	11.65ns
	1	44.60ns	1.12ns	4.39ns	53.8ns	11.71ns
	0.25	44.41ns	1.17ns	4.45ns	53.2ns	11.74ns

[a]Each value represents the mean of 3 samples. Triplicate determinations of each of 3 samples were made. Dry weight expressed as percent of fresh weight; other components, as percent of dry weight of the seeds.
*Significantly different from control at 0.05 level.
ns Not significantly different from control at 0.05 level.

Reproduced from Salunkhe, D. K., Wu, M. T., and Singh, B., *J. Am. Soc. Hortic. Sci.,* 96, 489, 1971a.

TABLE 15

Effects of Soil Treatment of *s*-Triazines on the Dry Weight and the Chemical Composition of Sweet Corn (2nd year) (Under Field Conditions)

Treatment

Compound	Concn. (1b/A)	Dry wt. %	Total N %	Soluble protein %	Starch %	Sugar %
Control		45.77[a]	1.19	4.55	53.12	14.25*
Simazine	4	46.03ns	1.29*	4.98*	49.20*	13.96*
	1	45.92ns	1.27*	4.91*	49.87*	14.03*
Atrazine	4	45.84ns	1.27*	4.80*	50.39*	14.12*
Prometone	4	45.97ns	1.29*	4.90*	50.63*	14.29*
Igran	4	45.64ns	1.27*	4.87*	49.45*	14.05*
Ametryne	4	45.71ns	1.28*	4.91*	50.13*	13.99*

[a]Each value represents the mean of 3 samples. Triplicate determination of each of 3 samples were made. Dry weight expressed as percent of fresh weight; other components, as percent of dry weight of the seeds.
*Significantly different from control at 0.05 level.
ns Not significantly different from control at 0.05 level.

Reproduced from Salunkhe, D. K., Wu, M. T., and Singh, B., *J. Am. Soc. Hortic. Sci.*, 96, 489, 1971a.

TABLE 16

Effects of Soil Treatment of Simazine, Prometone, Igran, and Ametryne on Amino Acids in the Seeds of Peas and Simazine on the Seeds of Sweet Corn (1st year) (Under Field Conditions)

	Peas							Sweet corn	
	Control	Simazine		Prometone		Igran	Ametryne	Control	Simazine
		0.5[a]	0.125[a]	0.5[a]	0.125[a]	0.5[a]	0.125[a]		4[a]
Alanine	1.13[b]	1.39	1.29	1.25	1.36	1.29	1.23	0.55	0.62
Valine	1.11	1.14	1.05	0.97	1.12	1.12	1.07	0.39	0.38
Glycine	0.98	1.21	1.07	1.12	1.17	1.10	1.08	0.26	0.34
Isoleucine	0.95	0.86	0.80	0.72	0.88	0.90	0.87	0.78	0.89
Leucine	1.76	2.17	1.99	1.96	2.04	1.99	1.92	0.66	0.70
Proline	0.86	1.05	0.97	0.98	1.02	0.94	0.96	0.24	0.28
Threonine	0.86	0.99	0.91	0.91	0.97	0.96	0.91	0.35	0.40
Serine	1.14	1.42	1.30	1.37	1.38	1.34	1.28	0.35	0.40
Methionine	0.21	0.26	0.14	0.18	0.23	0.18	0.22	0.18	0.20
Hydroxyproline	0.01	0.03	0.03	0.03	0.03	0.03	0.03	0.02	0.02
Phenylalanine	1.09	1.34	1.29	1.24	1.27	1.26	1.28	0.37	0.38
Aspartic acid	2.45	2.98	2.84	2.78	2.80	2.71	2.74	0.55	0.59
Glutamic acid	3.73	4.09	4.07	4.26	4.31	3.37	4.21	1.41	1.38
Tyrosine	0.56	0.77	0.64	0.66	0.69	0.64	0.68	0.25	0.27

TABLE 16 (continued)

Effects of Soil Treatment of Simazine, Prometone, Igran, and Ametryne on Amino Acids in the Seeds of Peas and Simazine on the Seeds of Sweet Corn (1st year) (Under Field Conditions)

		Peas						Sweet corn	
	Control	Simazine		Prometone		Igran	Ametryne	Control	Simazine
		0.5[a]	0.125[a]	0.5[a]	0.125[a]	0.5[a]	0.125[a]		4[a]
Lysine	1.66	2.02	1.90	1.90	1.82	1.78	1.76	0.25	0.32
Histidine	0.82	0.53	0.66	0.59	0.66	0.63	0.41	0.27	0.31
Arginine	1.45	1.63	1.81	1.49	1.20	1.68	1.62	0.23	0.22
Cystine/2	0.13	0.10	0.11	0.10	0.10	0.33	0.06	0.23	0.03
Total	20.90	23.98	22.87	22.51	23.05	22.25	22.33	7.02	7.57

[a]Pound/A of the chemical applied to the soil.
[b]Data are expressed as percent of dry weight.

Reproduced from Salunkhe, D. K., Wu, M. T., and Singh, B., *J. Am. Soc. Hortic. Sci.,* 96, 489, 1971a.

TABLE 17

Effects of Soil Treatment of Simazine, Prometone, Igran, and Ametryne on Amino Acids in the Seeds of Peas and Simazine on the Seeds of Sweet Corn (2nd year) (Under Field Conditions)

		Peas						Sweet corn	
	Control	Simazine		Prometone		Igran	Ametryne	Control	Simazine
		0.5[a]	0.125[a]	0.5[a]	0.125[a]	0.5[a]	0.125[a]		4.0[a]
Alanine	1.10[b]	1.35	1.27	1.23	1.34	1.29	1.18	0.55	0.61
Valine	1.09	1.12	1.12	1.00	1.11	1.11	1.06	0.36	0.37
Glycine	0.99	1.19	1.19	1.13	1.19	1.06	1.09	0.28	0.35
Isoleucine	0.96	0.88	0.88	0.70	0.89	0.91	0.90	0.22	0.23
Leucine	1.78	2.14	2.14	1.93	2.05	1.95	1.90	0.78	0.90
Proline	0.84	1.04	1.04	0.96	0.99	0.97	0.92	0.64	0.70
Threonine	0.86	0.97	0.97	0.90	0.94	0.93	0.88	0.26	0.28
Serine	1.13	1.36	1.36	1.38	1.34	1.28	1.27	0.34	0.38
Methionine	0.20	0.28	0.28	0.18	0.24	0.18	0.20	0.17	0.21
Hydroxyproline	0.01	0.03	0.03	0.03	0.03	0.03	0.03	0.02	0.02
Phenylalanine	1.08	1.35	1.35	1.23	1.30	1.25	1.35	0.39	0.41
Aspartic acid	2.44	2.95	2.80	2.80	2.76	2.69	2.78	0.56	0.60
Glutamic acid	3.69	4.13	4.06	4.25	4.33	3.45	4.31	1.43	1.40
Tyrosine	0.54	0.80	0.60	0.66	0.66	0.67	0.66	0.22	0.25
Lysine	1.66	2.00	1.86	1.88	1.80	1.81	1.75	0.24	0.26
Histidine	0.83	0.52	0.65	0.62	0.69	0.67	0.37	0.26	0.32
Arginine	1.47	1.65	1.84	1.51	1.21	1.64	1.67	0.23	0.24
Cystine/2	0.13	0.09	0.10	0.10	0.09	0.30	0.07	0.03	0.03
Tryptophan	0.11	0.13	0.13	0.13	0.13	0.15	0.11	0.03	0.04
Total	20.91	23.98	22.92	22.62	23.09	22.34	22.50	7.06	7.66

[a]Pound/A of the chemical applied to the soil.
[b]Data are expressed as percent of dry weight.

Reproduced from Salunkhe, D. K., Wu, M. T., and Singh, B., *J. Am. Soc. Hortic. Sci.,* 96, 489, 1971a.

TABLE 18

Effects of Foliar Treatment of Simazine and Prometone on Amino Acids in the Seeds of Peas and Sweet Corn[a] (Under Field Conditions)

Amino acid	Peas[b]			Sweet corn[c]		
	Control	Simazine 5 ppm	Prometone 2 ppm	Control	Simazine 5 ppm	Prometone 5 ppm
Alanine	1.10	1.48*	1.45*	0.55	0.65*	0.65*
Valine	1.09	1.18*	1.15*	0.36	0.41*	0.36
Glycine	0.99	1.21*	1.22*	0.28	0.36*	0.31*
Isoleucine	0.96	1.00	0.84*	0.22	0.25*	0.24*
Leucine	1.78	2.10*	2.01*	0.78	0.93*	0.82*
Proline	0.84	1.14*	1.06*	0.64	0.73*	0.70*
Threonine	0.86	1.03*	1.00*	0.26	0.31*	0.33*
Serine	1.13	1.42*	1.38*	0.34	0.40*	0.38*
Methionine	0.20	0.23*	0.19*	0.17	0.23*	0.17
Hydroxyproline	0.01	0.03*	0.04*	0.02	0.02	0.02
Phenylalanine	1.08	1.38*	1.25*	0.39	0.43*	0.46*
Aspartic acid	2.44	2.90*	2.87*	0.56	0.62*	0.62*
Glutamic acid	3.69	4.34*	4.20*	1.43	1.43	1.48*
Tyrosine	0.54	0.89*	0.81*	0.22	0.27*	0.28*
Lysine	1.66	2.05*	2.03*	0.24	0.34*	0.32*
Histidine	0.83	0.60*	0.68*	0.26	0.33*	0.32*
Arginine	1.47	1.68*	1.77*	0.23	0.26*	0.25*
Cystine/2	0.13	0.11*	0.10*	0.03	0.04*	0.04*
Tryptophan	0.11	0.17*	0.18*	0.03	0.05*	0.04*
Total	20.91	24.94*	24.23*	7.06	8.06*	7.78*

[a]Data are expressed as percent of dry weight.
[b]Treatments at 45 days after planting.
[c]Treatments at 60 days after planting.
*Significantly different from control at 0.05 level.

Reproduced from Singh, B., Vadhava, O. P., Wu, M. T., and Salunkhe, D. K., *J. Agric. Food Chem.*, 20, 1256, 1972c.

increased percentages of N compared with untreated controls. Total amounts of N per plant were significantly increased over the untreated controls by 18.9%. The treated plants contained significantly more nitrate-N, protein-N + nucleic acid-N, and chlorophyll-N + lipid-N than the untreated controls. The increases were 14.2, 39.0, and 31.2%, respectively. Total amounts of acid-soluble-N and ammonium-N were not significantly increased. There was an increase in amino acids that form the constituents common to nearly all protein molecules.

Effects of s-Triazines on Specific Enzymes of Plants

Increases in protein content of plants by subtoxic levels of s-triazines are well documented. Effects of s-triazines on biochemical events such as

the activity of enzymes relating to nitrogen and carbohydrate metabolism may help elucidate the mechanism of increase in protein contents. In their study on nitrate reductase activity in corn, Tweedy and Ries (1967) noted that the activity of this enzyme of leaf extract grown in the low level of nitrate and at the low temperatures increased almost tenfold after 7 days' exposure to Simazine. This activity increased in a linear fashion as the concentration of Simazine was increased up to 9.6 μM. Ebert and van Assche (1969) found that Atrazine at 10^{-6} to 10^{-8} M increased IAA-peroxidase activity. At 0.5×10^{-10} to $0.5 \times 10^{-21} M$, however, Atrazine depressed enzyme activity considerably.

Singh and Salunkhe (1970) demonstrated that foliar sprays of 0.5 ppm each of Atrazine, Simazine, Igran, and GS-14254 on bean plants increased the activity of nitrate reductase, ATPase, phosphorylase, transaminase, and amylase. Compared to the values determined for the control plants, the maximum stimulation of NRase activity was recorded on the 10th days for Simazine-, on the 5th and 10th day for Igran-, and on the 10th and 20th day for GS-14254-treated plants (Table 19). Significant stimulation of transaminase activity was evident at each sampling for all the treatments. The maximum increase in activity compared to the controls was noticed on the 10th day in all the treated plants (Table 20). All four treatments stimulated the activity of ATPase, with maximum stimulation occurring on the 5th day in almost every case (Table 21). Both enzymes of starch degradation, α-amylase and starch phosphorylase, were considerably stimulated by the treatments compared to the value in the control plants (Tables 22 and 23). The maximum increase in the activity of amylase and phosphorylase in Atrazine-treated plants was noted on the 10th day and in Igran- and GS-14254-treated plants on the 5th day. In Simazine-treated plants the maximum increase in amylase activity was on the 5th day; while in phosphorylase activity on the 10th day.

Millikan and Mann (1969) showed that Simazine (5 mg/kg soil) treatment increased the activity of catalase in corn. Brunetti et al. (1971) also

TABLE 19

Effects of Foliar Treatment of Atrazine, Simazine, Igran, and GS-14254 on the Activity of Nitrate Reductase in Leaves of Bush Bean Plants (Under Growth Chamber Conditions)

Days after treatment	Specific activity (mμ mole NO$_2$/mg protein/hr)				
	Control	Atrazine	Simazine	Igran	GS-14254
5	100.3	228.7**	139.8*	203.6**	170.3**
10	109.7	268.8**	249.8**	292.5**	242.8**
20	125.4	314.2**	217.7**	189.2**	287.1**

*Significantly different from control at 0.05 level.
**Significantly different from control at 0.01 level.

Reproduced from Singh, B. and Salunkhe, D. K., *Can. J. Bot.,* 48, 2213, 1970.

TABLE 20

Effects of Foliar Treatment of Atrazine, Simazine, Igran, and GS-14254 on the Activity of Glutamic-pyruvic Transaminase in Leaves of Bush Bean Plants (Under Growth Chamber Conditions)

Days after treatment	Specific activity (mμ mole pyruvic acid/mg protein/hr)				
	Control	Atrazine	Simazine	Igran	GS-14254
5	59.8	75.4*	142.1**	68.6*	69.7*
10	42.6	82.6**	104.4**	121.4**	72.4**
20	75.7	106.7**	130.4**	142.7**	99.1**

*Significantly different from control at 0.05 level.
**Significantly different from control at 0.01 level.

Reproduced from Singh, B. and Salunkhe, D. K., *Can. J. Bot.,* 48, 2213, 1970.

TABLE 21

Effects of Foliar Treatment of Atrazine, Simazine, Igran, and GS-14254 on the Activity of Adenosine Triphosphatase in Leaves of Bush Bean Plants (Under Growth Chamber Conditions)

Days after treatment	Specific activity ($m\mu$ mole PO_4/mg protein/hr)				
	Control	Atrazine	Simazine	Igran	GS-14254
5	30.4	72.1**	51.0**	50.7**	51.6**
10	99.8	140.0**	147.2**	165.4**	115.2*
20	77.6	133.7**	125.8**	103.1**	141.9**

*Significantly different from control at 0.05 level.
**Significantly different from control at 0.01 level.

Reproduced from Singh, B. and Salunkhe, D. K., *Can. J. Bot.,* 48, 2213, 1970.

TABLE 22

Effects of Foliar Treatment of Atrazine, Simazine, Igran, and GS-14254 on the Activity of Amylase in Leaves of Bush Bean Plants (Under Growth Chamber Conditions)

Days after treatment	Specific activity (mg starch hydrolyzed/mg protein/hr)				
	Control	Atrazine	Simazine	Igran	GS-14254
5	5.3	9.3**	11.9**	10.4**	11.7**
10	11.0	23.9**	14.5*	18.3**	17.7**
20	9.4	16.0**	13.8**	18.1**	16.6**

*Significantly different from control at 0.05 level.
**Significantly different from control at 0.01 level.

Reproduced from Singh, B. and Salunkhe, D. K., *Can. J. Bot.,* 48, 2213, 1970.

TABLE 23

Effects of Foliar Treatment of Atrazine, Simazine, Igran, and GS-14254 on the Activity of Starch Phosphorylase in Leaves of Bush Bean Plants (Under Growth Chamber Conditions)

Days after treatment	Specific activity ($m\mu$ mole Pi/mg protein/hr)				
	Control	Atrazine	Simazine	Igran	GS-14254
5	18.8	24.6**	29.3**	35.8**	51.3**
10	14.5	29.5**	39.4**	24.6**	26.9**
20	18.0	31.9**	40.5**	32.5**	36.9**

**Significantly different from control at 0.01 level.

Reproduced from Singh, B. and Salunkhe, D. K., *Can. J. Bot.,* 48, 2213, 1970.

found that Simazine, when present at low concentrations, increased N uptake and nitrate reductase activity in durum wheat seedlings grown in solution containing nitrate. At $0.5 - 1 \times 10^{-7}M$, Simazine apparently acts as a stabilizer of nitrate reductase, rather than as an inducer of new enzymes. Simazine also seemed to stimulate nitrate transport, indicating a structural or functional resemblance between nitrate reductase and the nitrate transport system.

Prometryne increased nitrate reductase activity, protein content, and nitrate content/fresh wt. in the sprouts of nodule-free plants in comparison with untreated controls. A growth-accelerating concentration of Prometryne increased the dry weight and absolute N content of the pods and seeds and also increased the number of ripe seeds per pod. Later, Wu et al. (1971b) demonstrated that foliar application of Simazine, Igran, and GS-14254 on 3-week-old pea seedlings caused an increase in the activity of δ-aminolevulinic acid dehydratase, an enzyme involved in the biosynthesis of porphyrin, heme compounds, and chlorophyll (Table 24). Wu et al. (1972a) also found that foliar sprays of Simazine, Propazine, Prometone, and Igran to the pea and sweet corn leaves increased the activity of NRase and transaminase, but not of RNase (Table 25).

TABLE 24

Effects of Foliar Treatment of Simazine, Igran, and GS-14254 on the Activity of Delta-aminolevulinic Acid Dehydratase of Pea Leaves (Under Greenhouse Conditions)

Days after treatment	Specific activity (μ moles PBG/mg protein/hr)			
	Control	Simazine	Igran	GS-14254
5	397	699**	681**	735**
10	405	661**	661**	620**
15	409	674**	674**	692**

**Significantly different from control at 0.01 level.

Reproduced from Wu, M. T., Singh, B., and Salunkhe, D. K., *Phytochemistry*, 10, 2025, 1971b.

TABLE 25

Effects of Foliar Treatment of *s*-triazines on Nitrate Reductase (NRase), Glutamate-pyruvate Transaminase, and Ribonuclease (RNase) Activity in the Leaves of 3-week-old Pea and Sweet Corn Seedlings 5 days After Treatment (Under Greenhouse Conditions)

Treatment		Pea			Sweet Corn		
		NRase	Transaminase	RNase	NRase	Transaminase	RNase
Compound	Concn. (ppm)	μmol NO_2/mg protein/hr	μmol pyruvate/mg protein/hr	μmol P_i/mg protein/hr	μmol NO_2/mg protein/hr	μmol pyruvate/mg protein/hr	μmol P_i/mg protein/hr
Control		52.6	36.8	96.4	24.8	21.8	50.6
Simazine	5	70.5**	54.2**	95.7ns	76.5**	24.9**	49.4ns
	2	64.8*	48.9*	98.3ns	54.3**	26.7*	50.3ns
Propazine	5	65.8*	46.6*	96.0ns	36.3*	33.7**	47.7ns
	2	69.4*	44.3*	91.5ns	41.5**	27.8*	46.3ns
Prometone	5	76.1**	69.3**	91.7ns	52.6**	31.5**	49.6ns
	2	67.3*	42.8*	95.5ns	47.4**	32.6**	48.7ns
Igran	5	83.0**	39.5*	90.1ns	42.7**	36.9**	51.1ns
	2	77.2**	41.8**	95.4ns	40.6**	30.7**	50.7ns

*Significantly different from control at 0.05 level.
**Significantly different from control at 0.01 level.
ns Not significantly different from control at 0.05 level.

Reproduced from Wu, M. T., Singh, B., and Salunkhe, D. K., *J. Exp. Bot.*, 23, 793, 1972a.

Foliar treatments of Simazine, Igran, or GS-14254 at 2 ppm also significantly stimulated the activity of starch phosphorylase (Figure 1), pyruvate kinase (Figure 2), cytochrome oxidase (Figure 3), and glutamate dehydrogenase (Figure 4) in the leaves of pea and sweet corn seedlings 5, 10 and 15 days after treatments (Wu et al., 1971c). In almost every case the maximum enzymic activity was noted on the 5th day in the s-triazine-treated plants. After that, enzymic activities decreased slowly but still remained higher than those of the control plants. When compared to the controls, the leaves of the Simazine-treated seedlings showed an increase of starch phosphorylase activity from 20 to 32% in pea and 27 to 61% in sweet corn; pyruvate kinase increased 25 to 45% in pea and 115 to 159% in sweet corn; cytochrome

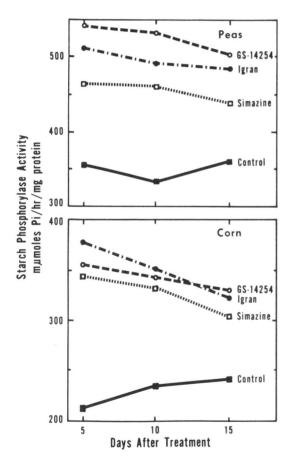

FIGURE 1. Effects of Simazine, Igran, and GS-14254 on the activity of starch phosphorylase in leaves of peas and sweet corn 5, 10, and 15 days after treatment. (Reproduced from Wu, M. T., Singh, B., and Salunkhe, D. K., *Plant Physiol.*, 48, 517, 1971c.)

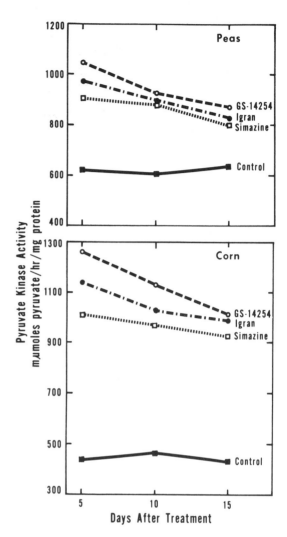

FIGURE 2. Effects of Simazine, Igran, and GS-14254 on the activity of pyruvate kinase in leaves of pea and sweet corn 5, 10, and 15 days after treatments. (Reproduced from Wu, M. T., Singh, B., and Salunkhe, D. K., *Plant Physiol.*, 48, 517, 1971c.)

oxidase, 45 to 73% in pea and 30 to 59% in sweet corn; and glutamate dehydrogenase, 27 to 54% in pea and 40 to 41% in sweet corn. In leaves of the Igran-treated seedlings the increase in the activity of starch phosphorylase was 33 to 44% in pea and 35 to 77% in sweet corn; pyruvate kinase increased 29 to 56% in pea and 101 to 130% in sweet corn; cytochrome oxidase, 39 to 89% in pea and 34 to 48% in sweet corn; and glutamate dehydrogenase, 26 to 67% in pea and 28 to 37% in sweet corn. In those of the GS-14254-treated seedlings the increase in the activity of starch phosphorylase was 37 to 52% in pea and 38 to 66% in sweet corn;

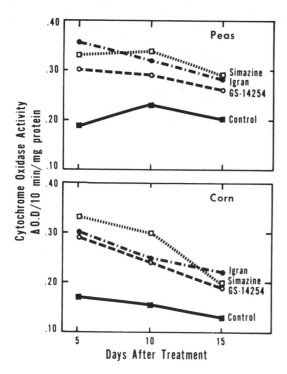

FIGURE 3. Effects of Simazine, Igran, and GS-14254 on the activity of cytochrome oxidase in leaves of peas and sweet corn 5, 10, and 15 days after treatments. (Reproduced from Wu, M. T., Singh, B., and Salunkhe, D. K., *Plant Physiol.*, 48, 517, 1971c.)

pyruvate kinase increased 36 to 67% in pea and 120 to 187% in sweet corn; cytochrome oxidase 26 to 57% in pea and 26 to 44% in sweet corn; and glutamate dehydrogenase, 22 to 50% in pea and 44 to 48% in sweet corn.

Effect of s-Triazines on Amino Acid Incorporation into Proteins

Rate of amino acid incorporation into protein is a parameter of protein synthesis. Wu et al. (1972b) reported that leaf discs from pea and sweet corn plants treated with Simazine, Propazine, Prometone, or Igran had higher rates of incorporation of L-leucine-U-^{14}C into soluble proteins than those from the controls (Table 26). They observed L-leucine-U-^{14}C incorporation rates by the leaf discs from s-triazine-treated plants appeared to be indicative of the rates of protein synthesis and enzyme activities in intact plants. This has been further substantiated by Pulver and Ries (1973), who showed that Simazine treatment increased ^{14}C-leucine incorporation in barley leaves from 10-day-old barley plants.

Electronmicroscopy: Protein bodies and Rough Endoplasmic Reticula

Besides chemical and biochemical evidences of subtoxic effects of s-triazines on proteins, any effects on the anatomy of target cells, especially organelles relating to protein synthesis and storage, would also be useful to elucidate the mechanism. Marked changes were noted in the ultrastructure of the cotyledon parenchyma cells of peas treated with Simazine (Wu et al., 1972b). The most obvious changes noted were in the shape and size of the protein bodies (Figures 5 and 6). The changes resembled those observed in this laboratory on bean cotyledons following applications of Atrazine, Simazine, Terbutryn, or GS-14254 (Figure 7) (Singh et al., 1972b). The cotyledons from the s-triazine-treated bean plants contained more protein bodies and more rough endoplasmic reticula. Treatments also increased the number of vesicles, which apparently contain protein, and the

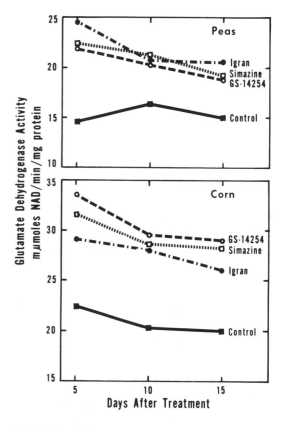

FIGURE 4. Effects of Simazine, Igran, and GS-14254 on the activity of glutamate dehydrogenase in leaves of peas and sweet corn 5, 10, and 15 days after treatment. (Reproduced from Wu, M. T., Singh, B., and Salunkhe, D. K., *Plant Physiol.*, 48, 517, 1971c.)

TABLE 26

Incorporation of L-leucine-U^{14}C into Protein in Leaf Discs of Peas and Sweet Corn 8 Days After Foliar Treatment with s-triazine Compounds (Under Greenhouse Conditions)

Compound	Concn. (ppm)	Pea Specific radioactivity (dpm/0.1 g discs)	Sweet corn Specific radioactivity (dpm/0.1 g discs)
Control		1024	434
Simazine	5	1237*	498*
	2	1194*	506*
Propazine	5	1208*	493*
	2	1222*	501*
Prometone	5	1231*	510*
	2	1229*	517*
Igran	5	1184*	503*
	2	1197*	509*

*Significantly different from control at 0.05 level.

Reproduced from Wu, M. T., Singh, B., and Salunkhe, D. K., *Experientia*, 28, 1002, 1972b.

amount of cytoplasmic ribosomes. In germinating seeds of *Vicia faba* the loss of an electron-dense material from the protein bodies was concurrent with a decrease in nitrogen in the cotyledon and almost certainly represented the transfer of protein to the growing apex (Briarty et al., 1970). In other words, a decrease or increase in the protein content of the cotyledons can be explained on the basis of the size and structural changes in the protein bodies.

Possible Mechanism(s) for Accumulation of Protein

A possible mechanism for the accumulation of protein in s-triazine-treated plants was proposed as shown in Figure 8 (Wu, 1971a). The increases in protein content that follow the application of s-triazines may result from an increased acitivity of nitrate reductase, glutamate dehydrogenase, and transaminase, producing more amino acids for protein synthesis. Stimulation of starch phosphorylase and α-amylase will increase starch degradation (thus reducing starch accumulation). This, coupled with stimulated pyruvate kinase, would give more substrates for amino acid synthesis. Besides, an increase in the activity of δ-aminolevulinic acid dehydratase and cytochrome

oxidase might accelerate the electron transport chain and consequently produce a great supply of ATP for the synthesis of protein.

Other Herbicides

As for other herbicides, Hiranpradit et al. (1972b) found that Bromacil® (5-bromo-3-sec-butyl-6-methyluracil) at 0.0225 ppm (0.086 μM) did not significantly affect soybean plant dry weights but it significantly increased both percentages and total amounts of N and P in plant shoots compared with untreated controls. The increases in total amounts of N and P per plant were up to 15.0 and 15.6% more than the untreated controls. The treated plants took up about 18% more nitrate-N than did the untreated controls, which resulted in significantly higher accumulation of protein N + nucleic acid-N in treated plants. Total amounts per plant of acid soluble-N and ammonium-N also tended to be increased. There was also an increase in amino acids. Fedtke (1972) showed that metabenzthi-azuron-treated wheat plants contained more water-soluble protein, amino acids, enlarged chloroplasts, and had an altered pigment composition with less reducing sugars over a period of 14 to 28 days after treatment.

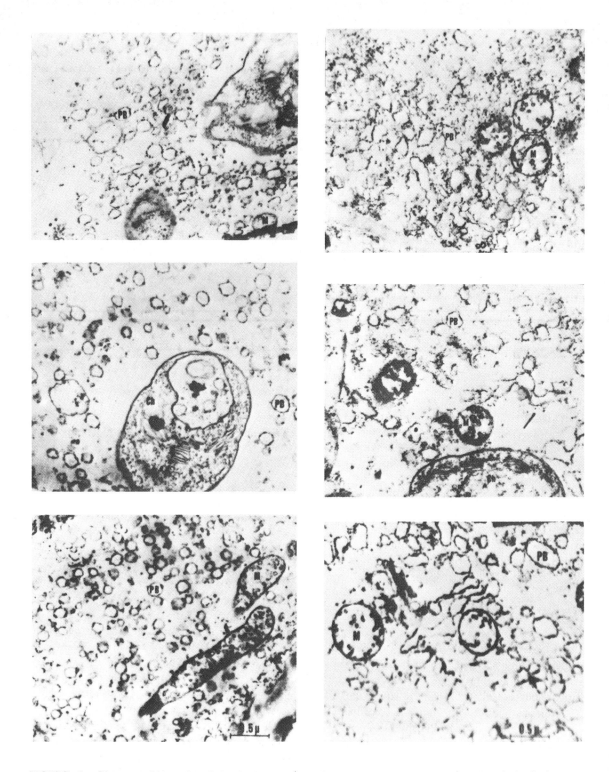

FIGURE 5. Electron micrographs of developing pea cotyledons of control plants. Note that the protein bodies are small and spherical in shape. PB = protein body; M = mitochondrion; and Ch = chloroplast. (Reproduced from Wu, M. T., Singh, B., and Salunkhe, D. K., *Experientia*, 28, 1002, 1972b.)

FIGURE 6. Electron micrographs of developing pea cotyledons of Simazine-treated plants. PB = protein body; M = mitochondrion; and Ch = chloroplast. (Reproduced from Wu, M. T., Singh, B., and Salunkhe, D. K., *Experientia*, 28, 1002, 1972b.)

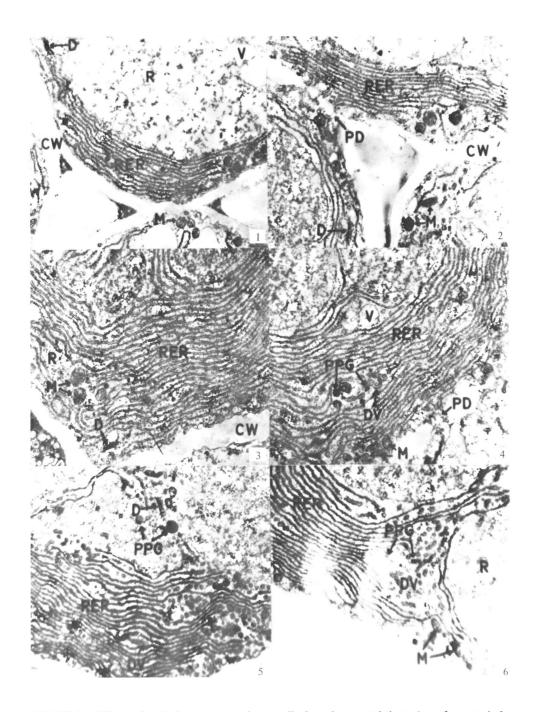

FIGURE 7. Effects of s-triazines on parenchyma cells from bean cotyledons. 1 to 2: control; 3: Atrazine-treated; 4: Simazine-treated; 5: terbutryn-treated; 6: GS-14254-treated. All magnification x 9,200. Rough endoplasmic reticulum (RER); presumptive protein granules (PG); cell wall (CW); distended visicle (DV); plasmodesma (PD); dictyosome (D); ribosome (R); vacuole (V); and mitochondria (M). (Reproduced from Singh, B., Campbell, W. F., and Salunkhe, D. K., *Am. J. Bot.,* 59, 568, 1972b.)

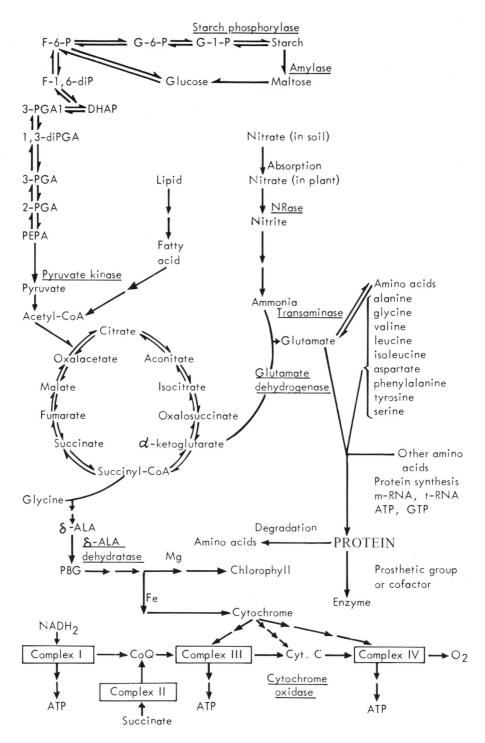

FIGURE 8. Proposed mechanism for the accumulation of more protein in s-triazine-treated plants. (Reproduced from Wu, M. T., Ph. D. Dissertation, Utah State University, Logan, Utah, 1971a.)

INCREASE IN CARBOHYDRATE CONTENT OF PLANTS BY CHEMICAL TREATMENTS

Naylor (1951) found maize plants treated with MH (maleic hydrazide) contained less glucose than controls, but up to 13 times more sucrose. Following the application of MH, there was a marked increase of both starch and sucrose in cotton plants (McIlrath, 1950) and a higher sucrose content in barley (Currier et al., 1951).

Payne et al. (1953) found that treatment of growing Red McClure potato plants with 2,4-D at a rate of ½ lb/A produced tubers with significantly higher protein content and a significantly higher specific gravity, which is indicative of a higher starch content. Significant decreases in reducing sugars at the time of harvest were shown by early combination treatments with MH and 2,4-D (Payne and Fults, 1955). They also found that late treatment of MH and 2,4-D alone and in combination significantly increased reducing sugars at the time of harvest. There was a significant increase in sucrose at the time of harvest in all the early treatments, with the exception of 2,4-D treatment. The sucrose yield of sugar cane was increased by 2,4-D application (Chacravarti et al., 1955).

Laboratory experiments with tomatoes, eggplants, and peppers sprayed with 2,4-D on the flower showed an increase in yield and sugar content (Yakushkina and Kravtsora, 1957). Kankanyan et al. (1971) showed that Prometryne and linuron increased the soluble sugars in carrots.

A study by Wu et al. (1970) indicated that soil fumigation with Telone® and Nemagon® brought about a significant increase in total sugars of carrot roots (Table 29). This was further confirmed by Salunkhe et al. (1971b) (Table 30), who also found that soil fumigation with Telone and Nemagon increased total sugar content of sweet corn seeds (Tables 31 and 32).

INCREASE IN VITAMIN CONTENT OF PLANTS BY CHEMICAL TREATMENTS

Luecke et al. (1949) found that application of 0.05 ml of 0.1% 2,4-D to red kidney bean increased the pantothenic acid in both stem and leaves. The effects of 2,4-D treatment on the vitamin C and carotene content of one-month-old buckwheat plants have been determined by Wort (1950). The herbicide was applied as a 100 or 1,000 ppm aqueous spray and also as doses of 10, 100, 1,000 μg per plant. Over a period of 8 days, the treatment of the 10 μg dose resulted in an increase of leaf ascorbic acid, but though the higher concentrations of 2,4-D produced an initial increase of vitamin C in the leaves, this was followed by a pronounced fall with the passage of time. Stem ascorbic acid was increased by all treatments. Carotene was reduced in both stem and leaves. Bean plants sprayed with 500 ppm 2,4-D showed lowered ascorbic acid content over an 11-day postspray period (Wort, 1950). The treatment of 400 ppm p-chlorophenoxy acetic acid on snap beans 4 days before picking retarded vitamin C degradation in the detached pods (Mitchell et al., 1949). After 4 days they were picked and the treated pods contained 40% more ascorbic acid than did the fruits from untreated plants.

Treatments with 10 ppm sprays of 2,4-D to pepper, tomato, and eggplant flowers increased the vitamin C content as well as dry weight (Yakushkina and Kravtsora, 1957). Fisher (1961) found that 2,4-D and 2,4,5-T treatment increased vitamin C content of tomato. Wort and Tathose (1967) later found that 2,4-D increased the ascorbic acid content of bush beans. Goren and Monselise (1966) showed improved ascorbic acid levels with Simazine and Atrazine applied to citrus trees. Urea and triazine treatments also increased ascorbic acid content of peaches (Belyaeva and Zavarzin, 1967). Chlorthiamid applications on four gooseberry cultivars proportionally increased their ascorbic acid contents (Byast and Hance, 1972). Kankanyan et al. (1961) found that linuron increased the ascorbate content of carrot roots by 30% and Prometryne by 18%.

Emerson et al. (1969) in a preliminary study found that Telone significantly increased the β-carotene content of carrot roots and sweet corn seeds. In an extensive study Wu et al. (1970) showed that soil fumigation with Telone and Nemagon brought about significant increases in the contents of total carotenes and β-carotene. The amounts of total carotene were 10 to 46% and of β-carotene 11 to 48% above the control values (Table 27).

Further study by Salunkhe et al. (1971b) indicated that soil fumigation with Telone and Nemagon brought about considerable increase in the contents of total carotenes (16 to 45%) and

109

TABLE 27

Effects of Soil Treatment of Telone and Nemagon on the Contents of Total and Reducing Sugars, Total Carotenes and β-carotene, and the Rate of Respiration of Carrot Roots (1st year) (Under Field Conditions)

Treatment						
Chemical	Dosage gal/acre	Total carotenes μg/g dry wt.	β-carotene μg/g dry wt.	Total sugars mg/g dry wt.	Reducing sugars mg/g dry wt.	Respiration $\mu 10_2$/g dry wt./hr
Control		503	473	431	172	1131
Telone	10	655**	617**	478**	152ns	853*
	20	643**	618**	658**	161ns	786*
	30	736**	704**	567**	154ns	741**
Nemagon	1	557**	526*	438ns	167ns	807*
	2	582**	540**	603**	175ns	732**
	3	645**	621**	600**	193ns	614**

*Significantly different from control at 0.05 level.
**Significantly different from control at 0.01 level.
ns Not significantly different from control at 0.05 level.

Reproduced from Wu, M., Singh, B., Wu, M. T., Salunkhe, D. K., and Dull, G. G., *Hort. Science,* 5, 221, 1970.

β-carotene (15 to 48%) in carrot roots (Table 28). Soil fumigation with these fumigants also influenced the composition of sweet corn seeds. Total carotenoids were increased up to 33% in 1969 and 26% in 1970 (Tables 29 and 30).

An ultrastructure study conducted by Wu and Salunkhe (1971b) showed that preplanting soil fumigation with Telone and Nemagon significantly influenced the chromoplasts of carrot roots when compared with those from the nonfumigated plots (Figures 9 to 11), especially the size, shape, and organization of the chromoplasts. The chromoplasts of the carrots grown in the fumigated soil were larger and contained more globules and crystals than those grown on the nonfumigated soil. The chromoplasts of the carrot roots from Telone-fumigated soil contained long needle crystals while those from Nemagon-fumigated soil contained short needle ones. The increase in amount and size of crystals and globules might result in an increase in carotene content of the carrot roots.

There are three possibilities which may cause changes in the amount of total carotenoids and β-carotene in plants grown in Telone-treated soil. First, Telone or one of its metabolites may cause an increase in the rate of biosynthesis of carotenoids in the plant; second, Telone or one of its

metabolites may decrease the rate of degradation of carotenoids in the plant; and third, Telone or one of its metabolites may be absorbed by the plant and further metabolized to take part directly in carotenoid biosynthesis. Berry et al. (1972) have indicated that 3-^{14}C-β-hydroxy-β- methyl glutaric acid (HMG) was effectively incorporated into isoprenoids by excised etiolated shoots as well as by the cell-free extracts of maize. The rate of incorporation indicated that HMG was not degraded to acetate or acetoacetate before entering the isoprenoid pathway. In absorption and translocation studies using bush bean plants, Berry (1971) showed that radioactive Telone enters the plant and that radioactivity appeared in almost every fraction of the plant 24 hr after treatment.

Berry et al. (1972) further demonstrated that Telone and 3-chloroallyl alcohol significantly inhibited the uptake of 3-^{14}C-HMG into carotenoids. The treatments with 3-chloro-1-propanol, 3-hydroxypropionic acid, and malonic acid at all dosage levels significantly decreased HMG uptake into carotenoids of the shoots, and only high levels of 3-chloroallyl alcohol and malonic acid inhibited the uptake of 2-^{14}C-MVA into carotenoids after 36 hr. They also found that Telone, 3-chloroallyl alcohol, *cis*-1,3-dichloropropene, trans-1,3-

TABLE 28

Effects of Soil Treatment of Telone and Nemagon on the Content of Total and Reducing Sugars, Total Carotenes and β-carotene, and the Rate of Respiration of Carrots (2nd year) (Under Field Conditions)

Treatment

Chemical	Dosage gal/acre	Total Carotenes µg/100 g	β-carotene µg/100 g	Total sugars g/100 g	Reducing sugars g/100 g	Total N g/100 g	Respiration $\mu 10_2$/hr/g
Control		5359	4927	4.70	2.27	0.14	108.8
Telone	10	6361**	5881**	5.95**	1.89ns	0.13ns	94.9*
	20	6746**	6270**	5.42**	2.04ns	0.14ns	88.5*
	30	7790**	7315**	5.72**	2.18ns	0.16ns	79.1**
Nemagon	1	6216**	5668**	5.00*	2.28ns	0.15ns	83.7*
	2	6734**	6182**	6.10**	1.98ns	0.14ns	82.8*
	3	7724**	6650**	5.95**	2.10ns	0.15ns	75.3**

*Significantly different from control at 0.05 level.
**Significantly different from control at 0.01 level.
ns Not significantly different from control at 0.05 level.

Reproduced from Salunkhe, D. K., Wu, M., Wu, M. T., and Singh, B., *J. Am. Soc. Hortic. Sci.*, 96, 357, 1971b.

TABLE 29

Effects of Soil Treatment of Telone and Nemagon on Total Carotenoids, Starch, Total Sugars, and Total N of Sweet Corn Seeds (1st year) (Under Field Conditions)

Treatment

Chemical	Dosage gal/acre	Total carotenoids µg/100 g	Starch g/100 g	Total sugars g/100 g	Total N g/100 g
Control		1459	21.4	3.81	0.51
Telone	10	1780*	21.3ns	3.50ns	0.51ns
	20	1882*	21.2ns	4.34*	0.52ns
	30	1947**	21.4ns	6.01**	0.52ns
Nemagon	1	1588*	21.6ns	4.13ns	0.51ns
	2	1690*	21.5ns	4.22ns	0.51ns
	3	1775*	21.3ns	4.96**	0.51ns

*Significantly different from control at 0.05 level.
**Significantly different from control at 0.01 level.
ns Not significantly different from control at 0.05 level.

Reproduced from Salunkhe, D. K., Wu, M., Wu, M. T., and Singh, B., *J. Am. Soc. Hortic. Sci.*, 96, 357, 1971b.

TABLE 30

Effects of Soil Treatment of Telone and Nemagon on the Contents of Total
Carotenoids, Starch, Total Sugar, and Total N of Sweet Corn Seeds (2nd year)
(Under Field Conditions)

Treatment

Chemical	Dosage gal/acre	Total carotenoids μg/100 g	Starch g/100 g	Total sugars g/100 g	Total N g/100 g
Control		1526	21.5	3.89	0.58
Telone	10	1720*	21.8ns	3.73ns	0.57ns
	20	1810*	21.2ns	4.43*	0.53ns
	30	1918**	21.8ns	5.59**	0.60ns
Nemagon	1	1652*	21.0ns	4.24*	0.58ns
	2	1703*	21.5ns	4.44*	0.57ns
	3	1821*	21.1ns	4.73**	0.55ns

*Significantly different from control at 0.05 level.
**Significantly different from control at 0.01 level.
ns Not significantly different from control at 0.05 level.

Reproduced from Salunkhe, D. K., Wu, M., Wu, M. T., and Singh, B., *J. Am. Soc. Hortic. Sci.,* 96, 357, 1971b.

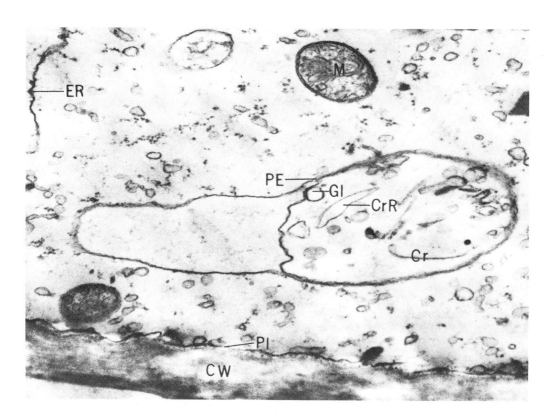

FIGURE 9. Electron micrographs of cells in the phloem of carrot roots grown in nonfumigated soil. Cr: pigment crystalloid; CrR: pigment crystalloid remnant; CW: cell wall; ER: endoplasmic reticulum; Gl: globule; M: mitochondrion; PE: plastic envelope; Pl: plasmalemma; and T: tonoplast. Mag. x 28,600. (Reproduced from Wu, M. and Salunkhe, D. K., *Experientia,* 27, 712, 1971b.)

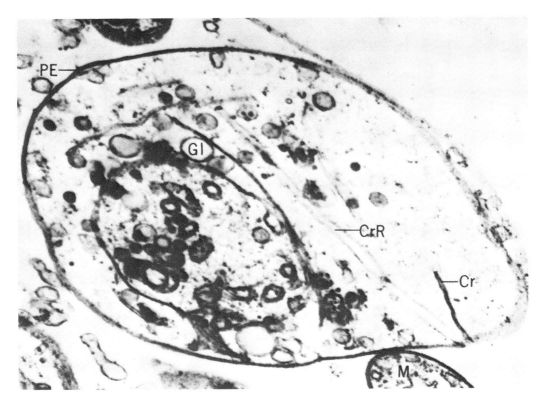

FIGURE 10. Electron micrographs of cells in the phloem of carrot roots grown in Telone-fumigated soil (30 gal/acre). Cr: pigment crystalloid; CrR: pigment crystalloid remnant; Gl: globule; M: mitochondrion; and PE: plastic envelope. Mag. x 28,600. (Reproduced from Wu, M. and Salunkhe, D. K., *Experientia*, 27, 712, 1971b.)

dichloropropene, 3-chloropropionic acid, 3-chloro-1-propanol, and malonic acid inhibited lipoxidase activity at low levels. The inhibitory concentrations of *cis*-1,3-dichloropropene and 3-chloroallyl alcohol are 100-fold less for equivalent inhibition of degradation than for the biosynthesis (i.e., 1.5×10^{-9} vs $2.0 \times 10^{-7} M$). The difference between these two treatment levels may result in a rate differential favoring long-term (2 to 3 months) increases in carotenoids. This would result from a greater suppression of lipoxidase activity at exposed concentrations within the treated plants (Berry, 1974).

Monson et al. (1971) in a study on the effect of nitrogen fertilization and Simazine on ·qualitative factors of the coastal Bermudagrass found that Simazine increased the carotene content by 22%. Kankanyan et al. (1971) found that Propazine, Prometryne, linuron, and chloramben increased the carotene content of carrot roots. Sweeney and Marsh (1971) reported that linuron and chloropropham increased the carotene content of

carrots, and that squash grown in soil treated with chloramben contained more carotene.

INCREASE IN MINERAL CONTENT OF PLANTS BY CHEMICAL TREATMENT

In his study on corn Freney (1965) noted that Simazine applied at 0.06 ppm in solution culture increased the uptake of nitrogen by 37%, phosphorus by 25%, magnesium by 24%, and potassium by 41%. Singh et al. (1972c) found that Atrazine, Simazine, or Terbutryn did not have any significant effect on the Ca, K, S, P, or Mg of bean pods grown under field or growth room conditions. However, the Fe content was significantly higher than in control plants in every case (Table 31). In the case of spinach, Atrazine, Simazine, and Terbutryn increased the amounts of K, Fe, P, and Mg. None of the compounds significantly influenced the Ca or S contents of spinach leaves (Table 32). Hiranpradit et al. (1972a) noted that

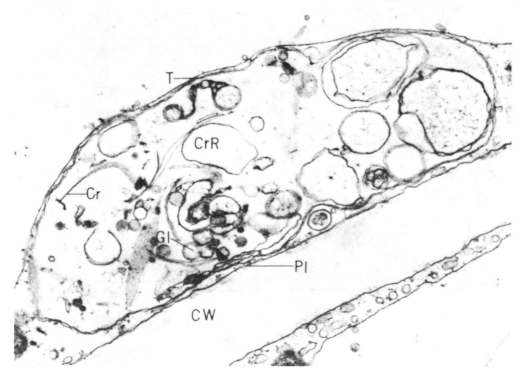

FIGURE 11. Electron micrographs of cells in the phloem of carrot roots grown in Nemagon-fumigated soil (3 gal/acre). Cr: pigment crystalloid; CrR: pigment crystalloid remnant; CW: cell wall; Gl: globule; Pl: plasmalemma; and T: tonoplast. Mag. x 28,600. (Reproduced from Wu, M. and Salunkhe, D. K., *Experientia*, 27, 712, 1971b.)

TABLE 31

Effects of Foliar Treatments of *s*-Triazines on Mineral Composition of Bush Bean Pods[a] (Under Field and Growth Chamber Conditions)

Crop	Treatment	Concn. (ppm)	Ca %	K %	S %	Fe mg/100 g	P %	Mg %
Beans under field conditions	Control	0	0.61	2.33	0.11	8.2	0.36	0.24
	Atrazine	0.5	0.58	2.45	0.12	9.8*	0.35	0.26
	Simazine	0.5	0.66	2.46	0.12	9.2*	0.39	0.26
	Terbutryn	0.5	0.68	2.37	0.13	9.2*	0.40	0.26
Beans under growth chamber conditions	Control	0	0.65	2.91	0.12	7.8	0.36	0.17
	Atrazine	0.5	0.67	2.87	0.13	9.2*	0.38	0.20
	Simazine	0.5	0.68	2.83	2.12	9.5*	0.35	0.18
	Terbutryn	0.5	0.67	2.79	0.12	9.8*	0.36	0.17

[a]Data expressed on dry weight basis.
*Significantly different from control at 0.05 level.

Reproduced from Singh, B., Vadhava, O. P., Wu, M. T., and Salunkhe, D. K., *J. Agric. Food Chem.*, 20, 1256, 1972c.

TABLE 32

Effects of Foliar Treatment of s-Triazines on Mineral Composition of Spinach Leaves[a] (Under Growth Chamber Conditions)

Treatment	Concn. (ppm)	Ca %	K %	S %	Fe mg/100 g	P %	Mg %
Control	0	1.94	0.49	0.39	15.2	0.37	0.80
Atrazine	1.0	1.83	0.61*	0.37	23.6*	0.41*	0.84*
Simazine	1.0	1.77	0.60*	0.38	19.8*	0.46*	0.85*
Terbutryn	1.0	1.69	0.55*	0.42	24.1*	0.44*	0.97*

[a]Data expressed on dry weight basis.
*Significantly different from control at 0.05 level.

Reproduced from Singh, B., Vadhava, O. P., Wu, M. T., and Salunkhe, D. K., *J. Agric. Food Chem.,* 20, 1256, 1972c.

Atrazine at 1.5 ppm significantly increased P, K, and Ca per plant over the untreated controls by 14.9, 16.4, and 13.8%, respectively.

INCREASE IN PIGMENT CONTENT OF PLANTS BY CHEMICAL TREATMENTS

Color of fresh fruits and vegetables is an important quality attribute which determines the consumer appeal of the produce. Any chemical treatment that can enhance color of the produce is highly desirable.

Bliss Triumph potatoes (Fults and Schaal, 1948; Ellis, 1949) and Red McClure potatoes (Fults and Payne, 1955), which produce red-skinned tubers, respond to treatment with 2,4-D by the production of tubers with a heightened red color. Treatment of the plants 69 days before harvest when the tubers were ¾ in. in diameter brought about the greatest color intensification. Similar effect has also been found by Payne et al. (1953) and Wort (1959). Wort (1962) later found that the foliar application of dusts containing 1.25% 2,4-D at the rate of 6 lb/acre to King Edward cultivar of potato (England) and to the Pontiac cultivar (Canada) has produced a similar intensification of the red color of the skins of these tubers.

The intensity of red coloration of fruit from Starking, Delicious Turley, and Red Rome cultivars of apple trees has been intensified by foliar sprays of fenoprop applied 2 to 4 weeks before normal harvest (Billerback et al., 1953). White (1953) also found that a foliar spray consisting of 25 ppm, 2,4,5-T plus 10 ppm fenoprop, applied 5 weeks before harvest, gave an increased red color to the inside of apples. Hoos et al. (1956) observed that 2,4,5-T sprayed on apricot trees at the pit-hardening stage of development hastened ripening and enhanced size, color, and yield of fruit.

A red pigment with a phenolic or quinoid structure was formed in the vascular tissue of sections of Jerusalem artichoke incubated on filter paper impregnated with solutions containing 10 mg/l of 2,4-D, 4-chloro-2-methylphenoxyacetic acid (MCPA), 2,4,5-T, and 2-methyl-3-chloro-phenoxyacetic acid (Swanson et al., 1956). *H. tuberosus, H. annuus,* and *Xanthium strumarium* treated with 2,4-D also produced the pigment. Malathion was found to enhance the extractable pigment content in cranberry fruits (Eck, 1968).

A treatment of Alar-85, 45 to 60 days before anticipated harvest, enhanced the color of apples (Batjer et al., 1964). Concentrations ranging from 2,000 to 4,000 ppm of Alar-85, or 6 to 12 lb/A, applied 2 weeks after bloom, advanced color of sour cherries (Unrath et al.,1969). It also advanced the color of sweet cherries (Kenworthy, 1970). A treatment of Alar-85 to the foliage at the beginning of pit hardening intensified the ground color of peaches (Bayers and Emerson, 1969).

RETENTION OF NUTRIENTS AND QUALITY IN PLANT PRODUCTS BY CHEMICAL TREATMENTS

Marth (1952) showed that growth regulators 2,4-D and 2,4,5-T have a pronounced retarding effect on the loss of the green color of broccoli and on cauliflower leaf abscission when applied in the field before or after harvest. Tuli et al. (1962) found that postharvest treatment of freshly harvested sweet cherries with N^6-benzyladenine resulted in more attractive and greener fruit pedicels with a higher chlorophyll content and less loss of fresh weight of the fruit during storage for 7 days at $20°C$. They also noted no beneficial effect in the retention of chlorophyll in strawberry fruit calyx as a result of such applications. Salunkhe et al. (1962) indicated that treatments of N^6-benzyladenine at concentrations of 5, 10, and 20 ppm did not increase the shelf life of sweet cherries, but the treatment did increase the shelf life of cauliflower, endives, parsley, snap beans, lettuce (head, leaf, and Romaine), radishes, bunching onions, and cabbage. Dedolph et al. (1961) explained such observed phenomena on the basis of respiration inhibition in asparagus spears. The same authors and also Gilbart (1962) found that, in addition to extending the shelf life and maintaining the appearance of freshly-harvested broccoli, a postharvest dip in N^6-benzyladenine at a concentration of 10 ppm also reduced weight loss, CO_2 evolution, and oxygen uptake under specified temperature conditions. On the other hand, Smock et al. (1962), Wittwer et al. (1962), and Wittwer and Dedolph (1963) suggested that such inhibition of respiration related to the stage of maturity when applications occurred. Moreover, Lipton and Ceponis (1962) observed a retardation of senescence and stimulation of oxygen consumption of kinin-treated lettuce tissues. The head lettuce sprayed with 10 ppm N^6-benzyladenine remained green longer than did the untreated lettuce. Zink (1961) found that N^6-benzyladenine applications on lettuce were essentially ineffective when applications were more than 3 to 4 days before harvest. The most satisfactory response occurred when the treatment was on the day of harvest. His findings suggest that the effect of this material disappears rapidly under field conditions. The same author attributed delayed senescence to maintenance of cell vigor and delayed proteolysis. Richmond and Lang (1957) reported that kinetin can reduce or prevent the accelerated protein loss that typifies detached leaves. At the same time, it delays the loss of chlorophyll and extends the life span of the leaf. They proposed that the chemical acts through its effect on the physiology of the tissue.

After a crop is harvested, general degradation sets in. This results in the destruction of soluble ribonucleic acid. Thus, protein synthesis slows down and as the mechanism of protein formation is hindered, the pigments and other constituents disintegrate. Salunkhe et al. (1962) explained that the primary step in the degradation of the soluble ribonucleic acid is thought to involve the loss of the end group adenine. Therefore, treatment with N^6-benzyladenine should provide the necessary adenine and restore the soluble ribonucleic acid molecules. Thus, protein synthesis would be maintained and the treated produce would stay fresh for a longer time.

Bessey (1960) found that lettuce heads treated with N^6-benzyladenine were maintained in fresh green condition considerably longer than untreated heads. Zink (1961) reported that N^6-benzyladenine acted as a senescence inhibitor and was effective in delaying the visual manifestation of senescence in lettuce, endive, Brussels sprouts, sprouting broccoli, mustard greens, radish tops, parsley, celery, green onions, and asparagus. Dedolph et al. (1963) and MacLean et al. (1963) investigated that pre- and postharvest treatments of N^6-benzyladenine reduced the respiration rate and extended the storage life of harvested broccoli.

El-Mansy (1964 and 1967) reported that pre- or postharvest treatments of lettuce with 6-furfurylaminopurine and N^6-benzyladenine followed by storage at $4.4°C$ and 85% R.H. resulted in higher values of moisture content, total chlorophyll, and total and insoluble nitrogen. Inhibition of CO_2 evolution during the storage was directly related to the concentrations of both 6-furfurylaminopurine and N^6-benzyladenine. On the other hand, the stimulation of O_2 consumption under the action of both chemicals was conversely related to their concentrations. The most promising results were obtained with postharvest treatments. 6-Furfurylaminopurine was more effective than N^6-benzyladenine and both maintained better quality ratings than characterized untreated control lettuce. These findings suggest the effectiveness of both chemicals on delaying the senescence of lettuce after harvest.

Halevy et al. (1966) found that senescence of "Grand Rapids" leaf lettuce was greatly reduced by postharvest treatment with Alar[®]. The deterioration and discoloration of mushrooms were inhibited by Alar, but Alar had little effect on broccoli and asparagus. They also observed that Alar actively helps preserve chlorophyll in leaves of bean and some other plants, and it increases the longevity of some perishable vegetables and mushrooms (Halevy and Wittwer, 1966). Cycocel® was effective in retarding senescence of lettuce and bean leaves, but was ineffective for broccoli and mushrooms (Halevy and Wittwer, 1966).

The preharvest foliar application of Randox® and the postharvest dips in N^6-benzyladenine and Randox solutions each reduced loss of sugar, concentration of raffinose, and rate of respiration of sugarbeet during storage (Wu et al., 1970). Soil fumigation of Telone and Nemagon decreased the respiration rates of carrot roots, thus increasing storage life (Wu et al., 1970; Wu, 1971a; and Salunkhe et al., 1971b).

CONCLUSIONS

World food production is lagging behind population growth despite all current national, bilateral, and international efforts to reverse this situation. Food that merely satisfies calorie needs is not enough. Adequate amounts of nutrients such as proteins, vitamins, and minerals are required for the normal maintenance of body tissues and functions, and additionally for growth, maturation, pregnancy, lactation, and recovery from injury and disease. Protein is particularly scarce and costly for the populations of developing countries. Today, over 300 million children, for lack of sufficient protein and calories, suffer grossly in physical growth and development, and many of these may also be impaired in mental development, learning, and behavior. Protein-calorie deficiencies also directly affect the health and economic productivity of adult populations. Any mineral or vitamin that is necessary for normal growth and metabolism can be presumed to influence the utilization of dietary protein. A deficiency of vitamins or minerals leads to failures of growth and deposition of protein in tissues. Vitamin A, iron, and calcium deficiencies have become severe in the U.S.A. in recent years.

The quality of products after harvest and during storage is also important, since it directly affects the available world food supply. Since 1940 agricultural revolutions in the technically advanced world have led to markedly increased supplies of wholesome food. Such improvement now is beginning to occur almost everywhere. This has come about primarily because of genetically improved cultivars and breeds, improved crop and livestock management practices, better plant and animal nutrition, farm mechanization, fertilization, irrigation, control of damaging weeds, diseases, insects and other pests and parasites.

This review paper presents results with and potential uses of certain chemicals to stimulate growth of plants and increase yield, to increase nutrient composition, and to increase nutrient retention and quality after harvest of economic plants or their products. Most of these findings, however, are of an experimental nature and practical applications have yet to be made. More research is needed in the search for more powerful chemicals and less expensive ways to use them. The side effects of these chemicals, especially the toxicological problems, also need to be explored.

ACKNOWLEDGMENTS

The studies conducted in our laboratories at the Utah State University and reported in this review were supported in part by research grant No. 12-14-100-9903-61 from the Human Nutrition Division, Agricultural Research Service, U.S. Department of Agriculture. The technical assistance of Drs. S. H. Lipton, G. G. Dull, B. Singh, D. L. Berry, Maureen Wu, Willis A. Gortner, Edith Weir, Mark Norris, John Ramsey, S. W. Dumford, H. M. LaBaron, S. Ichikawa, W. F. Campbell, J. O. Evans, G. D. Griffin, and C. M. Williams is gratefully acknowledged.

The authors wish to thank Professor S. K. Ries for the critical review of this manuscript.

REFERENCES

Alexander, A. G., *J. Agric. Univ. P. R.,* 53(3), 149, 1969.

Allinson, D. W. and Peters, R. A., *Agron. J.,* 62, 246, 1970.

Anderson, I. C., *Farm Chemicals,* 131(12), 29, 1968.

Barritt, B. H., *J. Am. Soc. Hortic. Sci.,* 95, 58, 1970.

Bartley, C. E., *Agric. Chem.,* 12, 113, 1957.

Bastin, R., Van Roey, G., and De Cat, W., *Meded. Fac. Landbouwwet. Rijksuniv. Gent.,* 35, 999, 1970; through *Chem. Abstr.,* 76, 42637, 1972.

Batjer, L. P., Williams, M. W., and Martin, G. C., *Proc. Am. Soc. Hortic. Sci.,* 85, 11, 1964.

Bayers, R. E. and Emerson, F. H., *J. Am. Soc. Hortic. Sci.,* 94, 641, 1969.

Belyaeva, T. V. and Zavarzin, V. I., *Khim. Selsk. Khoz.,* 5, 43, 1967; through *Chem. Abstr.,* 76, 136734m, 1972.

Berry, D. L., M.S. Thesis, Utah State University, Logan, Utah, 1971.

Berry, D. L., Singh, B., and Salunkhe, D. K., *Plant Cell Physiol.,* 13, 157, 1972.

Berry, D. L., Ph.D. Dissertation, Utah State University, Logan, Utah, 1974.

Bessey, P. M., *Univ. Ariz. Exp. Stn. Rep.,* 189, 5, 1960.

Billerback, F. W., Desrosier, N. W., and Tukey, R. B., *Proc. Am. Soc. Hortic. Sci.,* 61, 175, 1953.

Bodlaender, K. B. A. and Algra, S., *Eur. Potato J.,* 9, 242, 1966.

Briarty, L. G., Coult, D. A., and Boulter, D., *J. Exp. Bot.,* 21, 513, 1970.

Bruinsma, J., *Weed Res.,* 2, 73, 1962.

Bruinsma, J., *Plant Soil,* 24, 309, 1966.

Brunetti, N., Picciurro, G., and Ferrandi, L., *Agrochimica,* 16, 33, 1971; through *Chem. Abstr.,* 77, 713534, 1972.

Byast, T. H. and Hance, R. J., *Weed Res.,* 12, 272, 1972.

Chacravarti, A. S., Srivastova, D. P., and Khanna, K. L., *Sugar J.,* 18, 23, 1955.

Chapman, R. K. and Allen, T. C., *J. Econ. Entomol.,* 41, 616, 1948.

Claus, P., *Wein Wissenschaft Jahrgang,* 20, 314, 1965.

Colville, W. L., *Crops Soils Mag.,* April 8, 1969.

Coombe, B. G., *Nature (Lond.),* 205, 305, 1965.

Currier, H. B., Day, B. E., and Crafts, A. S., *Bot. Gaz.,* 112, 272, 1951.

Cyanamid International, *Cycocel, Plant Growth Regulant. Tech. Inf. Bull. FHT.,* Wayne, New Jersey, 1966.

Dedatta, S. K., Obcemea, W. M., and Jana, P. R., *Agron. J.,* 64, 785, 1972.

Dedolph, R. R., Wittwer, S. H., and Tuli, V., *Science,* 134, 1075, 1961.

Dedolph, R. R., Wittwer, S. H., Tuli, V., and Gilbart, D., *Plant Physiol.,* 37, 509, 1962.

Dedolph, R. R., Wittwer, S. H., and Larzedere, H. E., *Food Technol.,* 17, 1323, 1963.

DeVries, M. L., *Weeds,* 11, 220, 1963.

Deyton, E. D., M.S. Thesis, Michigan State University, East Lansing, Mich., 1973.

Eastin, E. F. and Davis, E. E., *Weeds,* 15, 306, 1967.

Ebert, E. and Van Assche, C. J., *Experientia (Basel),* 25, 758, 1969.

Eck, P., *HortScience,* 3, 70, 1968.

Ellis, N. K., *Down to Earth,* 4, 14, 1949.

El-Mansy, H. I., M.S. Thesis, Utah State University, Logan, Utah, 1964.

El-Mansy, H. I., Salunkhe, D. K., Hurst, R. L., and Walker, D. R., *Hortic. Res.,* 7, 81, 1967.

Emerson, G. A., Thomason, I. J., Paulus, A. O., Dull, G. G., and Snipes, J. W., *Symposium of the 8th International Nutritional Congress,* Prague, Czechoslovakia, Sept. 2, 1969.

Ennis, W. B., Jr., Jensen, L. L., Ellis, I. T., and Newson, L. D., in *The World Food Problem. A Report of the President's Science Advisory Committee,* Vol. III, 130, 1967.

Erickson, L. C., Seely, C. I., and Klages, K. H., *J. Am. Soc. Agron.,* 40, 659, 1948.

Fedtke, C., *Pest. Biochem. Physiol.,* 2, 312, 1972.

Finks, R. J. and Fletchall, O. H., *Weeds,* 15, 272, 1967.

Fisher, A. M., *Tr. Kaz. Inst. Epidemiol. Mikrobiol. Igigiemy,* 4, 565, 1961.

Food and Nutrition Board Recommended Dietary Allowances, 7th Ed., National Research Council, National Academy of Sciences, Washington, D.C., 1968.

Freney, J. R., *Aust. J. Agric. Res.,* 16, 257, 1965.

Fults, J. L. and Schaal, L. A., *Science,* 108, 411, 1948.

Fults, J. L. and Payne, M. G., *Am. Potato J.,* 32, 451, 1955.

Gast, A. and Grob, B., *Pest Technol.,* 3, 68, 1960.

Goren, R. and Monselise, S. P., *Weeds,* 14, 141, 1966.

Gramlich, J. V. and Davis, D. E., *Weeds,* 15, 157, 1967.

Gruzdev, G. S., Slovtsov, R. I., and Satarov, V. A., *Primen. Pestits. Sel. Khoz. Mater. Nauch. Konf.,* 33, 1969; through *Chem. Abstr.,* 75, 4375x, 1971.

Gvozdenko, T. M., *Dokl. Akad. Nark Azerb. S.S.R.,* 22, 55, 1966; through *Chem. Abstr.,* 67, 2312K, 1967.

Halevy, A. H., Dilley, D. R., and Wittwer, S. H., *Plant Physiol.,* 41, 1085, 1966.

Halevy, A. H. and Wittwer, S. H., *Proc. Am. Soc. Hortic. Sci.,* 88, 582, 1966.

Harrington, J. F., *Proc. Am. Soc. Hortic. Sci.,* 75, 476, 1960.

Hipp, B. W. and Cowley, W. R., *HortScience,* 4(4), 307, 1969.

Hiranpradit, H., Foy, C. L., and Shear, G. M., *Agron. J.,* 64, 267, 1972a.

Hiranpradit, H., Foy, C. L., and Shear, G. M., *Agron. J.,* 64, 274, 1972b.

Hoos, J. W., Leonard, S. J., and Luh, B. S., *Food Res.,* 21, 571, 1956.

Huffaker, R. C., Miller, M. D., Baghott, K. G., Smith, F. L., and Schaller, C. W., *Crop Sci.,* 7, 17, 1967.

Humphries, E. C. and Dyson, P. W., *Eur. Potato J.,* 10, 116, 1967.

Humphries, E. C., *Field Crops Abstr.,* 21, 91, 1968.

Humphries, E. C. and Pethigagoda, U., *Dtsch. Bot. Ges. (Neue Folge).* 3, 19, 1969.

Itakura, T., Kosaki, I., and Machida, Y., *Bull. Hortic. Res. Stn.,* (Ministry Agric. and Forest Series A, Hiratsuka), No. 4, 68, 1965.

Jaiswal, S. P., Saini, N. K., and Sharma, S. K., *Plant Soil,* 38, 33, 1973.

Kankanyan, A. G., Nazaryan, R. G., and Ananyam, A. M., *Isv. Sel'skokhoz. Nauk S.S.R.,* 14, 119, 1971; through *Chem. Abstr.,* 75, 75192M, 1971.

Kenworthy, A. L., *Proceedings of the 18th International Horticultural Congress (Abstr.),* Tel Aviv, 4, 1970.

Kesner, C. D. and Ries, S. K., *Weed Sci.,* 16, 55, 1968.

Khripunova, L. G., *Khim. Selsk. Khoz.,* 5, 764, 1967; through *Chem. Abstr.,* 68, 38443p., 1968.

Lipton, W. J. and Ceponis, M. J., *Proc. Am. Soc. Hortic. Sci.,* 81, 379, 1962.

Lorenzoni, G., *Estratto da Maydica,* 7, 115, 1962.

Luecke, R. W., Hamner, C. L., and Sell, H. M., *Plant Physiol.,* 24, 546, 1949.

MacLean, D. C., Dedolph, R. R., and Wittwer, S. H., *Proc. Am. Soc. Hortic. Sci.,* 83, 484, 1963.

Marth, P. C., *Proc. Am. Soc. Hortic. Sci.,* 60, 367, 1952.

Mazur, T. and Kasecka, T., *Zesz. Nauk. Wycz. Szk. Roln. Olsztynie,* 25, 925, 1969; through *Chem. Abstr.,* 73, 65275Y, 1970.

McIlrath, W. J., *Am. J. Bot.,* 37, 816, 1950.

McNeal, F. H., Hodgson, J. M., and Berg, M. A., *Can. J. Plant Sci.,* 4, 155, 1969.

Miller, M. D., Mikkelson, D. S., and Huffaker, R. C., *Crop Sci.,* 2, 114, 1962.

Miller, M. D. and Mikkelson, D. S., *Rice J.,* 73, 7, 1970.

Millikan, D. F. and Mann, D. R., *Physiol. Plant.,* 22, 1139, 1969.

Minshall, H., *Can. J. Bot.,* 38, 201, 1960.

Mitchell, J. W., Boyce, D. E., and Wilcox, M. S., *Science,* 109, 202, 1949.

Mohan Ram, H. Y. and Rustagi, P. N., *Agron. J.,* 61, 198, 1969.

Monson, W. G., Burton, G. W., Wilkinson, W. S., and Dumford, S. W., *Agron. J.,* 63, 928, 1971.

Morre, D. J., Rogers, B. J., and Gamble, R. C., *Phyton,* 22, 7, 1965.

Murneek, A. E., Wittwer, S. H., and Hemphill, D. D., *Proc. Am. Soc. Hortic. Sci.,* 45, 371, 1944.

National Academy of Science, Control of plant-parasitic nematodes, in *Principles of Plant and Animal Pest Control,* Vol. 4, 172, 1968.

Naylor, A. W., *Arch. Biochem. Biophys.,* 33, 340, 1951.

Nutman, F. J. and Roberts, F. M., *Trans. Br. Mycol. Soc.,* 45, 449, 1962.

Payne, M. G., Fults, J. L., and Hay, R. J., *Am. Potato J.,* 29, 142, 1952.

Payne, M. G., Fults, J. L., Hay, R. J., and Livingston, C. H., *Am. Potato J.,* 30, 46, 1953.

Payne, M. G. and Fults, J. L., *Am. Potato J.,* 32, 144, 1955.

Pulver, E. L. and Ries, S. K., *Weed Sci.,* 21, 233, 1973.

Richmond, A. E. and Lang, A., *Science,* 125, 650, 1957.

Ries, S. K., Larsen, R. P., and Kenworthy, A. L., *Weeds,* 11, 270, 1963.

Ries, S. K. and Gast, A., *Weeds,* 13, 272, 1965.

Ries, S. K., Chmiel, H., Dilley, D. R., and Filner, P., *Proc. Natl. Acad. Sci. (U.S.A.),* 58, 526, 1967.

Ries, S. K., Schweizer, C. J., and Chmiel, H., *BioScience,* 18, 205, 1968.

Ries, S. K., Moreno, O., Meggitt, W. F., Schweizer, C. J., and Ashkar, S. A., *Agron. J.,* 62, 746, 1970.

Salunkhe, D. K., Dhaliwal, A. S., and Boe, A. A., *Nature,* 195, 724, 1962.

Salunkhe, D. K., Wu, M. T., and Singh, B., *J. Am. Soc. Hortic. Sci.,* 96, 489, 1971a.

Salunkhe, D. K., Wu, M., Wu, M. T., and Singh, B., *J. Am. Soc. Hortic. Sci.,* 96, 357, 1971b.

Schulke, G., *Ber Schweiz. Bot. Ges.,* 80, 341, 1971; through *Chem. Abstr.,* 76, 818S, 1972.

Scrimshaw, N. S. and Gordon, J. E., *Malnutrition, Learning, and Behavior,* The M.I.T. Press, 1968.

Sell, H. M., Luecke, R. W., Taylor, B. M., and Hamner, C. L., *Plant Physiol.,* 24, 295, 1949.

Shafer, N. E., Cyanamid International, Wayne, N. J., 1967.

Shushu, G. E., *Fiziol. Biokhim. Genct. Rast. Kishinev.,* 4, 167, 1966; through *Chem. Abstr.,* 66, 104233g, 1968.

Singh, B. and Salunkhe, D. K., *Can. J. Bot.,* 48, 2213, 1970.

Singh, B., Salunkhe, D. K., and Lipton, S. H., *J. Hortic. Sci.,* 47, 441, 1972a.

Singh, B., Campbell, W. F., and Salunkhe, D. K., *Am. J. Bot.,* 59, 568, 1972b.

Singh, B., Vadhava, O. P., Wu, M. T., and Salunkhe, D. K., *J. Agric. Food Chem.,* 20, 1256, 1972c.

Smock, R. M., Martin, D., and Padfield, C. A., *Proc. Am. Soc. Hortic. Sci.,* 81, 51, 1962.

Sojkowski, Z., Staszewski, Z., and Sojkowska, T., *Hadowla. Rosl. Aklim. Nasiennictwo,* 13, 433, 1969; through *Chem. Abstr.,* 73, 129914, 1970.

Sosnovaya, O. M., *Fiziol. Biokhim. Kul't. Rast.,* 3, 44, 1971; through *Chem. Abstr.,* 75, 62378m, 1971.

Southam, C. M. and Ehrlich, J., *Phytopathology,* 33, 517, 1943.

Steenbjerg, F., Larsen, I., Jensen, I., and Belle, S., *Plant Soil,* 36, 475, 1972.

Steiger, R, von Primost, E., and Rittmeyer, G. R., *Z. Acker. Pflanzenbau.,* 130, 132, 1969.

Swanson, C. R., Hendricks, S. B., Toole, V. K., and Hagen, C. E., *Plant Physiol.,* 31, 315, 1956.

Sweeney, J. P., and Marsh, A. C., *J. Agric. Food Chem.,* 19, 854, 1971.

Swietochowski, B. and Glabiszewski, J., *Pamiet. Pulawski,* 28, 19, 1967; through *Chem. Abstr.,* 69, 1981g, 1968;

Tanimoto, T. and Nickell, L. G., *Report of the 25th Annual Meeting of the Hawiian Sugar Technologists,* 184, 1967.

Teubner, F. G. and Wittwer, S. H., *Science,* 122, 74, 1955.

Tognoni, F., Halevy, A. H., and Wittwer, S. H., *Planta (Berl.)* 72, 43, 1967.

Tompkins, D. R., *Proc. Am. Soc. Hortic. Sci.,* 87, 371, 1965.

Tompkins, D. R., *Proc. Am. Soc. Hortic. Sci.,* 89. 472, 1966.

Tuli, V., Dedolph, R. R., and Wittwer, S. H., *Mich. Agric. Exp. Stn. Q. Bull.,* 45(2), 223, 1962.

Tweedy, J. A. and Ries, S. K., *Plant Physiol.,* 42, 280, 1967.

Tweedy, J. A., Kern, A. D., Kapusta, G., and Millis, D. E., *Agron. J.,* 63, 216, 1971.

United States Government, *The World Food Problem. A report of the President's Science Advisory Committee,* Vol. I, II, and III, U.S. Government Printing Office, Washington, D.C., 1967.

Unrath, C. R., Kenworthy, A. L. and Bedford, C. L., *J. Am. Soc. Hortic. Sci.,* 94, 387, 1969.

Vergara, B. S., Miller, M., and Avelino, E., *Agron, J.,* 62, 269, 1970.

Weaver, R. J., *Proc. Am. Soc. Hortic. Sci.,* 61, 135, 1953.

Wedding, R. T., Kendrick, J. B., and Stewart, W. S., *Calif. Agric.,* 10, 4, 1956.

White, D. G., *Proc. Am. Soc. Hortic. Sci.,* 61, 180, 1953.

Wiedman, S. J. and Appleby, A. P., *Weed Res.,* 12, 65, 1972.

Wittwer, S. H. and Murneek, A. E., *Proc. Am. Soc. Hortic. Sci.,* 47, 285, 1946.

Wittwer, S. H. and Bukovac, M. J., *Mich. Agric. Exp. Stn. Q. Bull.,* 40, 352, 1957.

Wittwer, S. H., *Mich. Agric. Exp. Stn. Cir. Bull.,* 222, 1960.

Wittwer, S. H., Dedolph, R. R., Tuli, V., and Gilbart, D., *Proc. Am. Soc. Hortic. Sci.,* 80, 408, 1962.

Wittwer, S. H. and Dedolph, R. R., *Am. J. Bot.,* 50, 330, 1963.

Wittwer, S. H., *Proceedings of the 16th Annual Meeting of the Agricultural Research Institute,* Washington, D. C., 97, 1967.

Wolfenbarger, D. D., *J. Econ. Entomol.,* 41, 818, 1948.

Wort, D. J., *Western Canada Weed Control Conference ɩ'roceedings,* 4, 45, 1950.

Wort, D. J., *Proceedings of the 6th International Congress on Crop Protection,* Hamburg, 1, 497, 1959.

Wort, D. J., *World Review of Pest Control,* 1, 6, 1962.

Wort, D. J. and Tathose, V. S., *Abstract of the 6th International Congress on Plant Protection,* Vienna, 439, 1967.

Wu, M., Singh, B., Wu, M. T., Salunkhe, D. K., and Dull, G. G., *HortScience,* 5, 221, 1970.

Wu, M., M.S. Thesis, Utah State University, Logan, 1971a.

Wu, M. and Salunkhe, D. K., *Experientia,* 27, 712, 1971b.

Wu, M. T., Singh, B., Theurer, J. C., Olson, L. E., and Salunkhe, D. K., *J. Am. Soc. Sugar Beet Technol.,* 16, 117, 1970.

Wu, M. T., Ph.D. Dissertation, Utah State University, Logan, Utah, 1971a.

Wu, M. T., Singh, B., and Salunkhe, D. K., *Phytochemistry,* 10, 2025, 1971b.

Wu, M. T., Singh, B., and Salunkhe, D. K., *Plant Physiol.,* 48, 517, 1971c.

Wu, M. T., Singh, B., and Salunkhe, D. K., *J. Exp. Bot.,* 23, 793, 1972a.

Wu, M. T., Singh, B., and Salunkhe, D. K., *Experientia,* 28, 1002, 1972b.

Wyse, D. L., Ph.D. Dissertation, Michigan State University, East Lansing, Mich., 1974.

Yakushkina, N. I. and Kravtsora, B. E., *Dokl. Vses. (Ordena Lenina) Akad. S – Kh. Nauk. Im. V. I. Lenina.,* 22, 15, 1957; through *Chem. Abstr.,* 51, 140286, 1957.

Yasuda, A. K., Payne, M. G., and Fults, J. L., *Nature, (Lond.),* 176, 1029, 1956.

Zimmerman, P. W. and Hitchcock, A. E., *Proc. Am. Soc. Hortic. Sci.,* 45, 353, 1944.

Zink, F. W., *J. Agric. Food Chem.,* 9, 304, 1961.

DEVELOPMENTS IN TECHNOLOGY OF STORAGE AND HANDLING OF FRESH FRUITS AND VEGETABLES

Authors: **D. K. Salunkhe**
             M. T. Wu
             Utah State University
             Logan, Utah

INTRODUCTION

Modern technology has greatly increased the yields of fruits and vegetables. Higher yields are worth little, however, if the harvested crops are not consumed by people. One fourth of all produce harvested is not consumed because of spoilage between the time of harvest (during storage, transportation, and preprocessing holding) and time of purchase by the consumer. Billions of dollars are annually lost to growers, shippers, warehouse owners, and processors in the world. A sizable reduction also occurs in the nutritive values and general quality of much of the fresh produce and processed foods that do reach the consumer. In addition, such wastage is reflected in higher consumer prices. These losses are especially heavy in many parts of the world such as the developing countries that lack modern storage and processing facilities.

Producing areas tend to be remote from consuming areas; perishables such as fruits and vegetables therefore have to be harvested "green," and often remain in market channels for more than two weeks. Items in acutely short supply in one part of the U.S. may be in surplus in another part, and this situation also holds among nations.

Fruits and vegetables of Utah and the western U.S. take a transcontinental journey to the markets of the eastern U.S. and go even to those of Europe, the Middle East, Southeast Asia and South and Central America. More and more fruits and vegetables are being shipped wherever there is a demand. During our winter, Australian apples are brought into the U.S. Lettuce is shipped from the U.S. to Japan. Fresh strawberries are shipped to Germany and Hawaii from mainland U.S. Japan imports substantial amounts of bananas, tomatoes, and other fruits and vegetables from Taiwan.

Since fruits and vegetables are constantly subjected to spoilage caused by senescence and microbial decay while they remain in market channels, it is highly desirable to inhibit their ripening and senescence until they are to be consumed. Ideally we would like to be able to inhibit or accelerate the ripening process at will to insure maximum quality at a predetermined time. Whether to accelerate or delay the ripening of fruits and vegetables and whether to inhibit microbial growth on them depend upon consumer demand for the results of such activities. For example, certain fruits such as tomatoes and bananas are harvested in the green stage, packaged, and transported several thousands of miles. Before they are sold they are often treated with weak concentrations of ethylene or Ethephon to hasten ripening so that the consumer can buy them in a ripe condition. The storage life of fruits and vegetables can be prolonged days, weeks, or months by inhibiting ripening processes either with chemical treatments and/or with controlled atmosphere storage at lower temperatures until demand peaks.

The ripening and senescence of fruits and vegetables is a phenomenon long familiar to men. The chemical changes that occur in fruits and vegetables as they mature have been well characterized for many species of plants. The endogenous factors that control and regulate these changes, however, remain largely obscure. And the question of why any particular plant cell eventually dies is yet unanswered.

We do know that the ripening and senescence processes in fruits and vegetables involve the action of a group of chemical substances called plant hormones that are produced by the plants themselves. Men have synthesized chemicals that can function as the natural plant hormones do, and they have achieved some success in controlling the vital ripening process.

Fruits and vegetables, even when detached from trees or vines, are alive. Like all other living things, they carry on respiration. During respiratory processes, metabolites are oxidized in the presence of O_2 in the air. As a result, CO_2, water vapor, and heat are produced. The more rapid the rate of respiration after harvest, the more quickly produce ripens. Respiration rates or ripening of plant materials may be retarded by lowering the temperature, lowering the available O_2, or increasing the CO_2 concentration of the surrounding atmosphere. All three methods are applied in controlled atmosphere (CA) storage.

As yet, the complicated metabolic processes of fruits and vegetables are not completely understood. CA storage can influence many aspects· of the physiology and biochemistry of fruits and vegetables. The results can be beneficial or harmful, depending mainly on the products and the concentration of the atmosphere.

The use of a controlled atmosphere has varied effects on stored fruits and vegetables. In some cases the same atmosphere ·may damage one type of produce, while extending the shelf life of another type. The biochemical constituents of the fruits and vegetables are known to change in an unusual manner during CA storage, and substances normally not found in the fruit may appear.

CHEMICALS THAT DELAY RIPENING AND SENESCENCE

Researchers have synthesized compounds that function as natural plant hormones and that may be broadly classified as either hastening or delaying ripening. These include all categories of plant hormones, cytokinins, auxins, gibberellins, inhibitors, ethylene, absorbents, waxes, and other chemicals. Other chemicals are applied to inactivate microorganisms on highly perishable fruits and vegetables. These chemicals are fungicides, fungistats, and antibiotics.

Kinins and Kinetins

Applications of cytokinins, N^6-benzyladenine, and kinetin were previously conducted mainly on leaf tissues (Mothes et al., 1959; Pantastico, 1964; Richmond and Lang, 1957; Sugiura et al., 1962) and stem sections (Thimann and Manmohan, 1960). More recently these chemicals have been shown to delay chlorophyll degradation and senescence of leafy vegetables, spinach, peppers, beans, cucumbers, and other commodities. The gross effect is a retardation of yellowing by maintaining the protein level on the applied tissue. This effect becomes significant under prolonged storage or high temperatures. For instance, head lettuce sprayed with 10 ppm N^6-benzyladenine before harvest and held at a low temperature after harvest did not differ much from the controls (Lipton and Ceponis, 1962). Upon transfer to $20°C$, however, the untreated lots started to yellow, whereas N^6-benzyladenine-treated lettuce remained green.

Tuli et al. (1962) noted that postharvest treatment of freshly harvested sweet cherries with N^6-benzyladenine resulted in more attractive and greener fruit pedicels with a higher chlorophyll content and less loss of fresh weight of the fruit during storage for seven days at $20.1°C$. They also noted that there was no beneficial effect in the retention of chlorophyll in strawberry fruit calyx or cap as a result of such applications. Salunkhe et al. (1962) indicated that applications of N^6-benzyladenine at concentrations of 5, 10, and 20 ppm resulted in no increase of the shelf life of sweet cherries, but the application did increase the shelf life of cauliflower, endives, parsley, snap beans, lettuce (head, leaf, and Romaine), radishes, bunching onions, and cabbage. Similar results were obtained by Zink (1961) on Brussels sprouts, sprouting broccoli, mustard greens, radish tops, celery, green onions, and asparagus. Dedolph et al. (1961) explained such phenomena on the basis of respiration inhibition in asparagus spears. Dedolph et al. (1962) found that in addition to extending the shelf life and maintaining the appearance of freshly harvested broccoli, a postharvest dip in N^6-benzyladenine at a concentration of 10 ppm also reduced weight loss, CO_2 evolution, and oxygen uptake under specified temperature conditions. Smock et al. (1962), Wittwer et al. (1962), and Wittwer and Dedolph (1963) have further suggested that such inhibition of respiration is related to the stage of maturity when the chemical is applied. Moreover, Lipton and Ceponis (1962) observed a retardation of senescence and stimulation of oxygen consumption of kinin-treated lettuce tissues. The head lettuce that was sprayed with 10 ppm N^6-benzyladenine remained green longer than did the untreated lettuce.

Katsumi (1963) reported that both elongation and water uptake were inhibited in etiolated pea stem sections by kinetin. He observed that such inhibition was accompanied by decreased respiration. In an effort to find substances that overcome the kinetin inhibition, DNA, RNA, cysteine, methionine, arginine, uracil, ATP, and calcium were tested. None of these substances was effective in reversing kinetin inhibition of respiration.

Zink (1961) found that N^6-benzyladenine applications on lettuce were essentially ineffective when applications were made more than three to four days before harvest. The most satisfactory response occurred when application was on the day of harvest. His findings suggest that the effect of this material disappears rapidly under field conditions. The same author attributed delayed senescence to maintenance of cell vigor and delayed proteolysis. Richmond and Lang (1957) reported that kinetin can reduce or prevent the accelerated protein loss that typifies detached leaves. At the same time, it delays the loss of chlorophyll and extends the life of the leaf. They proposed that the chemical, not a biocide, acts through its effect on the physiology of the tissue.

After a crop is harvested, general degradation sets in. This results in the destruction of soluble ribonucleic acid. Thus, protein synthesis slows down and as the mechanism of protein formation is disturbed, the pigments and other constituents degrade. Salunkhe et al. (1962) explained that the primary step in the degradation of the soluble-type ribonucleic acid is thought to involve the loss of the end group adenine. A treatment with N^6-benzyladenine therefore should provide the necessary adenine and restore the soluble ribonucleic acid molecule. Thus protein synthesis would be maintained and the treated produce would stay fresh for a longer time.

El-Mansy (1964) and El-Mansy et al. (1967) reported that pre- or postharvest treatments of lettuce with kinetin and N^6-benzyladenine followed by storage at 4.4°C and 85% R.H. (relative humidity) resulted in higher values of moisture content, total chlorophyll, and total and insoluble nitrogen. Inhibition of CO_2 evolution during storage was directly related to the concentrations of both kinetin and N^6-benzyladenine, while the stimulation of O_2 consumption was conversely related to their concentrations. The most promising results were obtained with postharvest treatments. Kinetin was more effective than N^6-benzyladenine, but both maintained better quality ratings than the untreated lettuce. These findings suggest the effectiveness of both chemicals on delaying the senescence of lettuce after harvest. The postharvest dips in N^6-benzyladenine reduced loss of sugar, concentration of raffinose, and rate of respiration of the sugar beet during storage (Wu et al., 1970).

Cytokinins are sometimes combined with other growth regulators to enhance the effect of delaying senescence. Kaufman and Ringel (1961) observed that the delaying effect of N^6-benzyladenine on yellowing of cauliflower heads was more pronounced if combined with 2,4-dichlorophenoxyacetic acid. After 28 days' storage at 3.5°C, the control, N^6-benzyladenine, and 2,4-dichlorophenoxyacetic acid gave 11, 40, and 46% green leaves retained, respectively. A combination of 2,4-dichlorophenoxyacetic acid and N^6-benzyladenine produced 74% retention of green leaves.

Gibberellins

Postharvest treatment of gibberellic acid markedly retards ripening of tomatoes (Abdel-Kader et al., 1966; Dostal and Leopold, 1967), guavas (Saha, 1971; Teaotia et al., 1972), and bananas (Hsieh et al., 1967; Murata et al., 1965; Russo et al; 1968). The effect of ripening was indexed by lowered respiratory rate, retarded climacteric, and delayed color change. Abdel-Gawad and Romani (1968) did not observe such delay of ripening on pears. Based on tissue dice studies, they reported that this was due to inadequate penetrability.

Preharvest sprays of gibberellic acid were shown by Kitagawa et al. (1966) to have a striking effect in decreasing the rate of development, maturation, ripening, and senescence of kaki fruits. It is presently recommended in California for lemons and navel oranges to keep them on the tree beyond normal maturity. Some of the effects of applied gibberellic acid on citrus are retarded chlorophyll degradation, increased peel firmness, and delayed accumulation of carotenoids on navel oranges (Coggins and Hield, 1962; Coggins and Eaks, 1967; Eilati et al., 1969), higher soluble solids and ascorbic acid than in untreated "Libbon" lemons (Coggins et al., 1960), and no effect on fruit composition in "Valencia" oranges and grapefruits (Coggins et al., 1962). Lewis et al. (1967) explored the mechanism of action of gibberellic acid on navel oranges and found that treated fruits had a lower ratio of monovalent to divalent cations and a higher phosphorus level than the control. They proposed that the integrity of mitochondrial membranes was affected by gibberellic acid. A retardation of chlorophyll degradation may follow. Monselise and Goren (1965) observed an increase in peroxidase and catalase activity in the Shamouti orange of gibberellic acid-treated fruits. This also resulted in a reduction in both hesperidin and reducing sugars.

Auxins
Chlorophenoxyacetic Acid and β-Naphthoxyacetic Acid

Preharvest foliar applications of chlorophenoxyacetic acid and β-naphthoxyacetic acid at 25 ppm delayed the physio-chemical deterioration of "Coorg" mandarins in storage (Rodrigues and Subramanyam, 1966). A higher marketable condition of treated fruits after storage was due to reduced weight loss and retention of vitamin C. Similar reduction in weight loss and retention of vitamin C after harvest were observed on "Black Valentine" beans sprayed with chlorophenoxyacetic acid (50 to 500 ppm) 4 days before harvest. Pods from treated plants remained green even after all those from nontreated plants had shriveled and become dry and brittle (Steward, 1956).

Growth Retardants
Maleic Hydrazide (MH)

Many studies have been reported on the inhibit-

ing action of maleic hydrazide in storage sprouting of onions (Wittwer and Sharma, 1950; Wittwer and Paterson, 1951), radishes (Dewey and Wittwer, 1955), sugar beets (Wittwer and Hansen, 1951), turnips (Chourdri and Bhatnagar, 1955), carrots (Wittwer et al., 1950), and potatoes (Rao and Wittwer, 1955). Using a cultivar of onion with a short storage period (Hybrid Y-40), Patterson and Wittwer (1953) obtained 61% marketable bulbs in lots treated with 2,500 ppm maleic hydrazide compared to 38% of control after 6 months of storage. Moreover, reducing sugars of treated bulbs were retained. Similar sprout inhibition was obtained on onion cultivars by Celestino (1960) and Tabing and Gonzales (1956). Isenberg (1956) suggested that for a more effective sprout inhibition, maleic hydrazide should be applied when not more than 50% of the green tops are down. This is from 10 to 14 days before harvest. Thus, translocation of material is still active. Postharvest dips of onions in an aqueous solution gave the same control of sprouting, especially when the root plate of the bulbs was cut. However, the sprout inhibiting effect was offset by an increased amount of rotting (Acosta and York, 1957). Several other analogues of maleic hydrazide such as p-quinone, semi-carbazide, 2-thiouracil, phenylhydrazide, and isoicotinic acid hydrazide retarded sprouting of onion bulbs as a consequence of cessation of the growth of leaf primordia, with reduced protein synthesis and transaminase activity and respiration (Kato, 1970).

Irish potatoes sprayed 3 weeks before harvest with 3,000 ppm maleic hydrazide had considerably shorter sprouts than did the untreated tubers (Timm et al., 1959; Celestino and Ventura, 1960). Moreover, treated potato tubers held in storage for 7 months at 7° and 13°C were higher in specific gravity than untreated ones (Salunkhe et al., 1953). Sprouting during storage of sweet potatoes was also reduced by spraying the crops with 5,000 ppm maleic hydrazide 2 weeks after harvest (Gooding and Campbell, 1964).

The effect of maleic hydrazide on the ripening process varies with different fruits and may depend on time of application and amount of chemical absorbed. Harvested mangoes dipped in 1,000 and 2,000 ppm maleic hydrazide in water showed delayed ripening but it was less effective in hot than in cold water (Krishnamurthy and Subramanyam, 1970). Southwick and Lachman (1953) obtained a similar delaying effect of maleic hydrazide on ripening of tomatoes. The chemical counteracted the stimulatory effect of water on ripening.

Maleic hydrazide applied on sapota, however, hastened ripening (Lakshminarayana and Subramanyam, 1967), increased the respiration rate, and induced a high transpiration rate. Overcash (1955) and Crandall (1955) failed to influence the ripening of apples applied with maleic hydrazide as foliar spray one to six weeks before harvest, but increased flesh firmness of fruits during storage. Again, as in other work on growth regulators, more detailed investigations on the reactions of various fruit tissues to the chemical are needed.

N,N-Dimethyl Aminosuccinamic Acid (Alar)

Increased fruit firmness, better fruit color, and early maturation are some of the effects of Alar when applied before harvest (Edgerton and Hoffman, 1966; Kenworthy, 1970; Mattus, 1968; Tehrani and Tibor, 1969). Subramanyam and Sebastian (1970) also reported that 5-min dips of "Alphonso" mango in aqueous solutions of 2,500 ppm Alar at $25°C$ had no effect on the carotene content and ripening of fruits. However, when hot ($53°C$) aqueous solutions of Alar were used, respiration, carotene level, and ripening were enhanced compared to hot water alone. Subramanyam and Sebastian (1970) suggested that hot water increases the permeability of the external tissue and activates the carotene-synthesizing enzymes.

Patil et al. (1971a) reported that Alar and Ethrel (Ethephon) were significantly effective in inhibiting the synthesis of solanine in potatoes under storage conditions in supermarkets. Studies of Dennis et al. (1965) and Ryugo and Sachs (1969) revealed that Alar depressed the incorporation of mevalonic acid into isoprenoids. Perhaps Alar inhibits the synthesis of solanine by inhibiting the conversion of mevalonic acid into isoprenoid moiety of solanine.

2-Chloroethyltrimethyl Ammonium Chloride (CCC or Cycocel)

Cycocel effectively retarded senescence of some vegetables. Shelf life of leaf lettuce was doubled and deterioration in quality of broccoli heads and asparagus spears was retarded by immersion of the cut stem bases or dipping the vegetables in solutions of CCC (Halevy and Wittwer, 1965). Cycocel did not, however, delay senescence of mushrooms (Halevy and Wittwer, 1966) or control chlorophyll and solanine formation in potato tubers (Patil et al., 1971a).

Isopropyl N-(3-chlorophenyl) Carbamate (CIPC), Isopropyl n-Phenyl Carbamate (IPC), and Methyl Ester of α-Naphthalene Acetic Acid (MENA)

Vaporized CIPC has given excellent sprout control on potatoes for up to a year in $10°C$ storage at a dosage as low as 1 g/4 bushels, whereas the less potent MENA required 1 g/bushel (Sawyer and Dallyn, 1957).

Foliar sprays of IPC reduced the rate of respiration, ripening, and spoilage of sapota fruits during storage (Lakshminarayan et al., 1967). Ascorbic acid was retained although there was a slight reduction in the total sugar content of sapota fruits treated with 100 ppm of IPC.

Metabolic Inhibitors
Cycloheximide and Actinomycin-D

Frenkel et al. (1969) reported the cycloheximide prevented ripening of pear fruits when administered at the early preclimacteric stage (unripe), but it was progressively less effective at the later ripening stages. Flesh softening, chlorophyll degradation, and ethylene synthesis were severely inhibited in the pears by cycloheximide treatment. When cycloheximide was applied to pears that had begun to ripen, the ripening was slowed. Fruits at the nearly ripe stage treated with cycloheximide ripened to the same extent as the controls. Similar effects on ripening were also observed on preclimacteric fruits treated with Actinomycin-D. Ethylene treatment did not reverse the complete inhibition of ripening imposed by Actinomycin-D.

Vitamin K

Vitamin K_1 (2-methyl-3-phytyl-1,4-naphthoquinone) or vitamin K_3 (2-methyl-1,4-naphthoquinone) has been demonstrated by Beccari (1969) to inhibit banana ripening at temperatures as high as $60°C$. Beccari proposed that the action of vitamin K acts as a general metabolic block, probably by acting as an inhibitor of electron transfer during ethylene production.

Maleic Acid

Early works on metabolic inhibitors were carried out by Copisarow (1936a) who claimed that maleic acid in amyl acetate retarded the ripening of Jamaica bananas, pineapples, citrus, and other deciduous fruits. This was later disputed by Isaac (1937, 1938). He reported that the delaying effect of maleic acid on ripening was only transient. Moreover, amyl acetate had a detrimental effect on fruit skin, leading to susceptibility to fungal infection. This may not be practical as plant sprays against fungi attack as has been claimed by Copisarow (1936b).

Ethylene Oxide

Mangoes treated with ethylene oxide also showed a definite delay in ripening. Moreover, ethylene oxide-treated fruits developed an attractive skin color and a firm texture. Inhibition of ripening due to ethylene oxide treatment was also observed by Lieberman and Mapson (1962), Lieberman et al. (1964), and Ayres et al. (1964) on tomatoes. Lieberman and Mapson (1962) reported that treatment of green tomato fruits in an atmosphere containing 0.75% ethylene oxide for approximately 20 hr delayed ripening and ethylene production by 5 to 21 days. Larger concentrations of ethylene oxide or longer times of exposure prevented ripening. Tomato fruits that had started ripening and were on the climacteric rise, producing relatively large amounts of ethylene and having some red color, were not appreciably retarded by ethylene oxide treatment. It was suggested that ethylene oxide may be an endogenous metabolic antagonist of ethylene and that the interaction between the two may determine the degree of ripening attained at any stage.

Sodium Dehydroacetic Acid (Na-DHA)

Watada and Scott (1970) found that NaDHA (0.5%) delayed the ripening and reduced the rate of respiration of strawberry fruits. They proposed that Na-DHA inhibits the respiratory enzyme activity.

Carbon Monoxide

Carbon monoxide flushing of mushrooms in CA and at $3.3^{\circ}C$ extended storage life fourfold (Besser and Kanner, 1970). If, however, ethylene oxide is mixed with the carbon monoxide, the color and texture of mushrooms are not preserved. Apparently, the two chemicals are competitive at their site of action.

Ethylene Absorbents

Attainment of maximum possible storage life is the goal of storage studies. Usually, treatments such as waxing, low O_2, high CO_2, and ripening inhibitors are combined. However, even when using the optimum levels of each treatment for ripening inhibition, endogenous ethylene is always a problem. Thus, hypobaric storage was tried to keep ethylene below the threshold level (Burg and Burg, 1966). Many chemical formulations have been tried with a similar purpose. Brominated activated carbon (Chiang, 1968; Smock, 1967) and celite with $KMnO_4$ (Forsyth et al., 1967) were tried on different fruits with varying results. Scott et al. (1968) finally developed a more practical material, $KMnO_4$ on vermiculite. Later (Scott et al., 1970) they showed that bananas in film bags containing potassium permanganate (to absorb ethylene) were firmer than fruits in sealed bags with $Ca(OH)_2$. This effect was more marked after 38 days. It was estimated that about two weeks of additional storage life was obtained by packing potassium permanganate with the fruit.

A commercial preparation of the absorbent called "Purafil" (alkaline potassium permanganate on a silicate carrier) was produced by Carbon Chemical Company, West Virginia, and proved effective in the complete absorption of ethylene from bananas held in sealed polyethylene bags (Liu, 1970). Thus, ethylene absorbents in conjunction with CA storage could have a notable commercial potential in the future.

Fumigants
Methyl Formate and Methyl Bromide

Fumigants to control fungi growth have also been examined to see if they inhibit ripening of mangoes. Subramanyam et al. (1969) found that methyl formate, aside from delaying spoilage, gave maximum inhibition of ripening throughout the storage period of 20 days at the ambient temperature. Methyl bromide gave less inhibition of mango ripening than ethylene oxide or methyl formate, but the effect was still marked in comparison with untreated fruit (Subramanyam et al., 1969). Preplanting soil fumigation with Telone and Nemagon decreased the respiration rate of carrot roots, thus increasing storage life (Wu et al., 1970; Salunkhe et al., 1971b).

CHEMICALS THAT HASTEN RIPENING AND SENESCENCE

Ethylene and Ethylene-releasing Compounds
Ethylene

Ethylene is a volatile substance emanated by

fruits and vegetables that is a specific ripening hormone. An exogenous application of minute quantities (about 1 ppm) of the gas stimulated the respiratory rate, hastened ripening, and inhibited germination of seeds, sprouting of potatoes, and epinasty of leaves. The response to ethylene by climacteric fruits is noted only during preclimacteric phase, whereas in nonclimacteric fruits respiratory activity and ripening could be accelerated in all stages of maturity. This differential response to ethylene by climacteric and nonclimacteric fruits can be attributed to their relative abilities to produce ethylene endogenously. If ethylene is produced in sufficient quantity, external application of ethylene would not produce a response. However, additional ethylene application in nonclimacteric fruit might be effective because of the low rates of ethylene production.

Biogenesis of Ethylene

Several substrates have been reported capable of being converted to ethylene in model systems; however, some of them failed to produce ethylene in vivo. Substances proposed as possible precursors included glucose, linolenic acid, and methionine.

Mechanism(s) of Ethylene Action

According to Lyons and Pratt (1964), ethylene could induce changes in permeability of the mitochondrial membranes, thus facilitating the increased movement of ATP and thereby initiating several reactions including increased rates of respiration. They further observed swelling of mitochondria of pear tissue during ripening following ethylene treatment.

Ethylene-releasing Compounds

An important breakthrough in ethylene effects and physiology was the chemical synthesis of 2-chloroethylphosphonic acid (Ethephon) by Kabachnik and Rossiiskaya (1946) and the report of Maynard and Swan (1963) that ethylene is released from Ethephon. All biological effects attributed to ethylene gas are duplicated by aqueous solutions of Ethephon. Gas-confining chambers needed for ethylene application are unnecessary. Applications of Ethephon have since been attempted to affect various plant processes.

Several other ethylene-releasing compounds were formulated recently. Foremost among them is 2-(4-chlorophenylthio-)-triethylamine hydrochloride (CPTA). However, most of the published works are on Ethephon.

Tomatoes

The yield from a once-over harvest of tomatoes was markedly increased when Ethephon at 250 to 2,000 ppm concentrations was applied any time from pollination to a mature-green stage (Angell, 1970; Dostal and Wilcox, 1970; Garrison, 1968; Iwahori et al., 1968; Iwahori and Lyons, 1969, 1970; Klater and Rudich, 1970; Rabinowitch et al., 1970; Robinson et al., 1968; Sims, 1969). Treatments higher than 2,000 ppm of Ethephon may result in a high percentage of fruit damage by sunscald (Sims, 1969). This may be prevented, however, by applying diatom talc or other opaque-white materials (Klater and Rudich, 1970). Dennis et al. (1970), Gull (1969) and Salunkhe et al. (1971a) observed that ripening was stimulated by Ethephon. Thus, once ripening had commenced or when 50 to 55% of the fruits were ripe (Dostal and Wilcox, 1970), Ethephon treatment did not enhance usable fruit yields. Garrison (1968) also observed that Ethephon had little effect on the time to incipient coloring when application was made 40 days after pollination, soon after the mature-green stage but just before the first sign of ripening. All cultivars that normally develop lycopene turned red after treatment with 1,000 ppm Ethephon (Robinson et al., 1968). They also noted that *Lycopersicon peruvianum* (green-fruited species) and *L. esculentum* mutant, neither of which synthesizes lycopene, did not turn red with Ethephon. Apparently Ethephon and ethylene do not induce better color or quality development than that of untreated tomato fruit. However, maximum color development is attained much earlier. Ethephon penetrates the cuticle of the fruit and hastens color development of the treated part (Robinson et al., 1968).

To summarize, foliar or postharvest applications of Ethephon on tomatoes would have the following advantages: (a) sorting cost of tomatoes may be reduced due to uniformity of ripening, (b) fast ripening rates will reduce weight loss and may prolong shelf life, (c) ripening rooms are not necessary, (d) increase of yield from once-over harvest, and (e) maturity may be hastened early in the season to obtain marketable fruit with premium prices.

Other Fruits

Foliar sprays of Ethephon also produced uniform ripening and concentrated total crop maturity on other fruits. Pineapples matured uniformly with foliar sprays of 0.5 to 1.0 lb Ethephon per acre 1 or 2 weeks before harvest (Audinay, 1970; O'Brien, 1970). Cantaloupes treated with a 1,000 ppm foliar spray of Ethephon just before the first harvest ripened earlier than unsprayed melons (Kasmire et al., 1970; Tyler et al., 1970). Moreover, the total yield of melons after Ethephon treatment was greater than that of the controls (Kasmire et al., 1970). An isolated case of a 3- to 10-day delay in fruit maturity was obtained by Karchi (1970) in melons treated with Ethephon. According to Karchi, flower initiation was delayed by Ethephon. This is contrary to the results obtained by Dutcher (1971). Flower initiation of mangoes was actually induced by Ethephon treatment. Figs treated with 250 to 500 ppm Ethephon ripened 2 to 3 weeks earlier than the controls (Crane et al., 1970). Chile and pimiento peppers ripened earlier when sprayed with 250 to 1,200 ppm Ethephon at their chocolate brown stage (Brendler, 1970; Sims et al., 1970). As a preharvest spray, therefore, Ethephon is very useful to concentrate crop maturity, especially in fruits with normally prolonged harvest periods, thereby making picking and processing more economical.

As a postharvest treatment, Ethephon also promotes ripening of many fruits. "Lakatan" bananas dipped for 5 min in 2,500 ppm Ethephon solution attained their climacteric peak 5 days earlier than did untreated fruits (Bondad, 1971). The rate of ripening (indexed by chlorophyll loss, softening, and respiratory activity) of banana fruits treated with 1,000 ppm Ethephon was found to be the same as those treated with 100 ppm ethylene gas (Russo et al., 1968).

Citrus fruits degreened with Ethephon (1,000 ppm for 3 sec to 10 min) were comparable to those obtained in commercial type degreening rooms (Daito and Hirose, 1970; Fuchs and Cohen, 1969; Oberbacher, 1968; Young et al., 1970). Degreening of "Valencia" oranges treated with Ethephon was somewhat slower in the open than in the dark (Oberbacher, 1968). Chauhan and Parmar (1972) noted that treated "Mosambi" oranges stored in the dark were brighter yellow and glossier than those kept in the open. Gloss and shelf life of "Ladu" mandarins and "Valencia" oranges were further improved by "Prima-fresh" wax treatment after complete degreening with Ethephon (Pantastico, unpublished data). Degreening temperature of Ethephon-treated citrus was

more efficient at 25°C than at 17°C and was inhibited at 6°C. In some citrus varieties Ethephon actually induced carotenoid synthesis, aside from hastening chlorophyll degradation (Daito and Hirose, 1970).

Abscisin

The role of abscisin in the ripening process of fruits and vegetables is not clear at this time. The application of abscisin, however, does accelerate senescence in detached fruits and leaves. Cooper and Henry (1971) reported that an abscisic acid treatment of oranges accelerated the breakdown of chlorophyll and increased the synthesis of carotenoids. Ripening of banana slices was also accelerated by abscisic acid (Vendrell, 1970).

Ascorbic Acid and β-Hydroxyethyl Hydrazine (BOH)

Ascorbic acid, Cu-EDTA (Cupric ethylenediaminetetracetate) and BOH induced ethylene formation in tangerines (Cooper et al., 1968; Palmer et al., 1967; Rasmussen and Cooper, 1968) when sprayed before harvest. However, when tried as a postharvest dip, none of these chemicals produced a degreening rate similar to that induced by ethylene on lemons and "Satsuma" oranges (Tsai and Chiang, 1970), although degreening was slightly better than in untreated citrus fruits.

Acetylene, Propylene, and Calcium Carbide (CaC_2)

Calcium carbide, which generates acetylene, is also used to hasten fruit ripening. Calcium carbide was used on bananas (Ancheta, 1957; Salas, 1957) and citrus fruits (Tsai and Chiang, 1970). Tomatoes are apparently resistant to CaC_2 treatment. Much higher concentrations than were used for bananas were required for tomato ripening (Bondad et al., 1971; Bondad and Pantastico, 1971). For banana ripening, ethylene was 100 times more effective than acetylene (Pantastico and Mendoza, 1970). Smoke from any burning material such as leaves, twigs, or straw may also hasten ripening. Gas chromatographic analysis of such smoke showed that its active ingredients were ethylene and acetylene (Pantastico and Mendoza, 1970). For degreening lemons and "Satsuma" oranges (Tsai and Chiang, 1970), a postharvest treatment with 500 ppm acetylene for 2 days was less effective than a similar treatment with 10 ppm ethylene, but still produced marked degreening after 1 week (57% and 49% for lemons; 56% and 47% for "Satsuma," respectively). According to them, 1,000 ppm of acetylene are needed to obtain the same rate of degreening as with less ethylene, confirming the data obtained by Pantastico and Mendoza on bananas.

Propylene, like ethylene, promotes fruit ripening, inhibits elongation of pea subapical sections, and causes epinasty (Crocker et al., 1935). Burg and Burg (1967) determined the molecular requirements for ethylene action and found propylene to be the next most active compound to show ethylene-like action. The equivalent concentration of propylene to cause a half-maximum ethylene response was 130 times that of ethylene. McMurchie et al. (1972) found that propylene induced ethylene production in bananas but not in citrus fruits. They therefore suggested that two ethylene-producing systems exist in climacteric fruits, and that the one responsible for autolysis is missing in nonclimacteric fruits. A study on apple fruits by Sfakiotakis and Dilley (1973) supported the idea of two ethylene-producing systems in climacteric fruits. They also indicated that the nonautocatalytic system is limited by O_2 supply, while the autocatalytic system is less dependent upon O_2, which indicates the difference in affinity for O_2 of the enzymes involved in the two ethylene-producing systems.

Alcohols

Studies by Rakitin et al. (1957) suggested that additions of methanol induced a ripening response in persimmon fruits. Boe (1966) found that ethanol injections caused differential ripening of tomato fruit. The degree of ripening was inversely proportional to the amount of ethanol injected. He also reported that among alcohols such as ethanol, hexanol, octanol, dodecanol, heptanol, and hexadecanol, which are used in dip treatments of green tomatoes prior to ripening, hexanol immersions seemed to accelerate the ripening of the tomato fruit.

Fatty Acids

Oleification has been known as a practice to hasten fig fruit maturation since the 3rd century B.C. Why and how the oil treatments stimulate maturation was not clear, however, until recently. Hirai et al. (1966, 1967a, 1967b, 1967c, 1967d, 1968) conducted a series of studies designed to elucidate the physiological mechanism of oleification. They applied various kinds of oils and related chemicals to the eye of fig fruits. In a day or two, fruits treated with any of several vegetable oils began to increase in size, reaching full color and maturity within a few days. Untreated figs remained green and hard. The effect of animal (whale) oil on maturation was somewhat less than that of vegetable oil. Mineral oil was inferior to vegetable and animal oils in results (Hirai et al., 1966). Oil treatment induced respiration and ethylene production and influenced the activities of growth substances. These results suggested that oleification accelerates the normal course of ripening in fig fruits (Hirai et al., 1967c).

Linoleic acid also strikingly stimulated the maturation of figs. Further, fatty acid oxidation products were found to be an effective inducer. Among these, the ripening effect of acetaldehyde was the most conspicuous. Ethylene and ethanol also showed the effect to a considerable degree (Hirai et al., 1967d).

From these results it was concluded that the effect of oleification may be due to high enzymatic activity in the cells of fruits induced by the volatile substances such as the acetaldehyde of ethylene, which is derived from both fatty acid oxidation and sugar decomposition (Hirai et al., 1968).

It should be pointed out that although the specific metabolic site affected by the chemicals used to promote ripening remains obscure, the effects may be connected with stress phenomena. High chemical concentrations, elevated temperatures, or mechanical injury may lead to increased metabolic activity. Production of ethylene can thus be enhanced, leading to ripening. This is why hot water treatment (Singh et al., 1969) hastened the ripening of tomatoes.

CHEMICALS THAT MAY HASTEN OR DELAY RIPENING AND SENESCENCE

2,4-Dichlorophenoxyacetic Acid (2,4-D), 2,4,5-Trichlorophenoxyacetic Acid (2,4,5-T), and 2,4,5-Trichlorophenoxypropionic Acid (2,4,5-TP)

Murata et al. (1965) studied the effects of growth regulators on the ripening of "Shinzun" bananas and found that their climacteric ascent and respiration were hastened by 100 to 1,000 ppm 2,4-D. This was substantiated by Shah and

Ghosh (1972) with "Kabuli" bananas, by Stewart (1956) with "Fortuna" and "Guatemala" bananas. Shah and Ghosh (1972) reported, however, that although the period of ripening was shortened by applications of 2,4-D and 2,4,5-T on fully mature fruits, a greenish-yellow color was produced instead of the typical yellowish-green tinge. On the other hand, a 100 ppm concentration of 2,4,5-TP delayed ripening of "Pairi" mangoes and did not improve the skin color (Krishnamurthy and Subramanyam, 1970). These variations in the effects of 2,4-D, 2,4,5-T, and 2,4,5-TP on various ripening processes caused different results of application trials. For instance, stimulation of ripening and maturation process of apples was reported by Lott and Rice (1955), Teskey and Francis (1954), Southwick (1953), and Thompson (1952), but other workers (Batjer et al., 1954; Bullock and Rogers, 1952; Haller, 1954; Hatton, 1955) have not always noted this effect. Storage life of citrus fruits was prolonged by applications of 2,4-D and 2,4,5-T (Stewart, 1949, 1956; Stewart et al., 1952), mainly due to a delayed degreening (Hatton, 1958; Hield and McCarthy, 1956). There was less physiological loss in weight, less vitamin C loss, and a high marketable percentage in stored 2,4,5-T-treated "Coorg" mandarin oranges (Rodrigues and Subramanyam, 1966). Damigella (1962) reported that 2,4,5-T was better than 2,4-D in promoting early ripening of mandarins. Using "Nagpur" and "Darjeeling" mandarin oranges, however, Lodh et al. (1963) showed that 2,4,5-T was in fact less effective than 2,4-D in prolonging the storage life of the fruit. But even a 1 ppm concentration of 2,4,5-T has a noticeable effect in delaying senescence of pineapple fruit (Gortner, 1969; Gortner and Leeper, 1969). In fact, a treatment of pineapples with a 100 ppm solution of 2,4,5-T is a process patented by the Pineapple Research Institute of Hawaii (Gortner, 1963). This process extended the shelf life of pineapples for several days by delaying senescence. On the other hand, ripening of peaches (Weinberger, 1951) and prunes (Zielinski, 1951) was hastened by low 2,4,5-T levels on peaches. A 30 to 40 ppm concentration was injurious to peaches (Higdon, 1951). Fruit ripening was hastened when 2,4-D (100 ppm), or 2,4,5-T (100 ppm), or 2,4,5-TP (25 ppm) was sprayed on the fruit of sapota trees (Lakshminarayana et al., 1967). Ripening of guava was also markedly hastened by 2,4-D and 2,4,5-T

(Saha, 1971). Rate of ripening was doubled in guavas treated with 200 ppm 2,4,5-T. In "Black Mission" fig fruits, ripening was accelerated by 17 days by a water spray of 10 to 25 ppm 2,4,5-T. However, considerable leaf damage also occurred (Stewart, 1956). The same dose of this chemical applied on "Calimyra" figs accelerated ripening by 30 to 60 days, but leaf damage was still noted. Maxie and Crane (1967) reported that 2,4,5-T indirectly accelerated ripening of fig fruit by stimulating ethylene synthesis. They noticed that the stage of fruit maturity at the time of auxin application had a pronounced effect on the response. Abeles and Rubenstein (1964) found that auxin stimulated the synthesis of ethylene by tomato fruit and this action is also dependent on the stage of maturity at the time of treatment. Treated immature tomato fruits produced more ethylene than did fruits treated at the pink stage. This could be the reason why treatment of 2,4-D had little effect on the respiration of mature green tomatoes (Emmert and Southwick, 1953).

An application of 2,4-D, especially if combined with BA, retarded yellowing of cauliflower (Kaufman and Ringel, 1961) and other green vegetables (Zink, 1961). Leaf drop in stored cauliflower and cabbage has been reduced by an application of 100 to 500 ppm 2,4-D at 1 to 7 days before harvest (Stewart, 1956). Apparently, therefore, there is no confusion in the effect of 2,4-D and 2,4,5-T on green color retention. The effects of these two auxins on ripening would depend on the concentrations used, the species and varieties of fruits and vegetables, preharvest factors, and specific experimental conditions. Cappelini and Fideghelli (1970) consistently demonstrated that 2,4,5-TP application promoted ripening in peaches but delayed it in table-grapes. Analyses of fruits showed that peroxidase and catalase activities of peaches and grapes differed considerably in response to 2,4,5-TP treatment. Oxidative reactions were retarded in table-grapes and enhanced in peaches.

Indole-acetic Acid (IAA) and Naphthalene Acetic Acid (NAA)

Naphthalene acetic acid has been used more extensively in application studies than IAA because it is cheap and readily available aside from the fact that their biological actions do not differ much. Indole-acetic acid is used only on a laboratory scale and on works dealing with metabolic phenomena. Thus, it was shown by Lampe (1971) that IAA inhibits polygalacturonase activity in the early stages of tomato ripening, but not later. A similar delaying effect on ripening was obtained with NAA by Lampe (1971) on tomatoes, by Stewart (1956) on "Smooth Cayenne" pineapples, and by Soni et al. (1972) on "Basrai Dwarf" bananas. After 16 days of storage at 20.5° to 29°C, the bananas treated with 100 ppm NAA and 4% wax emulsion had lower reducing, nonreducing, and total sugars, acidity, and total soluble solids compared with the controls (Soni et al., 1972). Storage life of onion bulbs was also prolonged when a 0.4 ppm NAA was used as a foliar spray (Chhongkar and Sengupta, 1972). Cauliflower and cabbage wrapped in shredded paper containing 50 to 100 mg NAA and stored at 0°C showed a reduction in leaf abscission and weight loss (Stewart, 1956).

Preharvest sprays of NAA on apples and peas were reported (Allen, 1953; McKenzie, 1953; Padfield, 1949) to accelerate ripening and shorten life. However, Gerhardt and Allmendinger (1945) and Marshall et al. (1948) found neither an increased ripening nor a deteriorated storage quality on fruits sprayed with NAA. Batjer and Moon (1945) and Smock et al. (1954) partly resolved the conflicting results when they reported that NAA stimulated ripening of certain summer apple varieties but had little or no such effects on later ripening cultivars.

Auxin-ethylene Interactions on Ripening

Stewart (1956) had demonstrated interactions between auxins and ethylene. He showed that when 2,4-D was applied prior to an ethylene treatment the accelerating effect of ethylene on ripening was reduced. However, no such inhibition was obtained if ethylene was applied first. Hormonal interaction is doubtful here because one was applied after the other. Effective interrelations commonly require a joint application of chemicals.

In general, the effect of auxins on ripening is that of retardation rather than acceleration. Ripening was hastened mostly in cases of high auxin concentrations with resultant metabolic damage. Another reason for the observed early induction of ripening with auxin treatment is the inability of auxin solutions to penetrate the entire fruit tissue. Auxins penetrating only the epidermal cells would enhance the production of ethylene which in turn can initiate ripening of tissues devoid of applied auxins. A thorough auxin penetration should therefore be assured. Thus, vacumm infiltration of IAA and 2,4-D made by Frenkel and Dyck (1973) showed that ethylene, even at elevated levels, hardly overcame the inhibitory effects of IAA and 2,4-D on ripening. This overriding influence of auxin on ethylene was also observed by Vendrell (1969) in banana peel slices and Sacher (1967) in bean pericarp segments.

If it is accepted that auxins function as resistance factors in ripening (Mapson, 1970), then inactivation of auxin action during ripening is a prerequisite before ethylene can act on the tissue (Frenkel, 1972). Auxins may be degraded through increased IAA oxidase activity (Frenkel, 1972). Hall and Morgan (1964) even suggested that ethylene promotes the activity of IAA oxidase. Lieberman and Kunishi (1972) further stated that ethylene not only influences the destruction of IAA but also suppresses its synthesis and blocks its site of action. Moreover, ethylene may have inhibitory effects similar to those of gibberellic acid and cytokinins. If these observations are true, then the resistance to ripening of fruits at their early stages of maturity or of attached avocados may be due to the action of auxins and other plant hormones at inhibitory levels. Another possible source of inhibition, closely related to those of senescence retardants, is the presence of enzyme inhibitors.

CHEMICALS THAT CONTROL POSTHARVEST MICROBIAL GROWTH

The serious economic nature of postharvest diseases is evident from the fact that the cost of processing and marketing most fruits and vegetables greatly exceeds the value of the raw materials. Decay losses of a few percent often turn an otherwise successful agricultural venture into an economic failure despite months of the most meticulous farming practices. A major portion of losses of fresh fruits and vegetables during shipping and marketing is attributed to diseases caused by fungi and bacteria.

The most serious postharvest diseases are those that cause rapid and extensive breakdown of high-moisture commodities, sometimes spoiling the entire package and causing secondary infections in the advanced stages of the diseases. This condition is exemplified by attacks of *Rhizopus* sp. on stone fruits and strawberries, *Penicillium* sp.

on citrus and pome fruits (blue and green mold decays), and *Erwinia carotovora* on leafy vegetables and potatoes (bacterial soft rots). It is estimated that 30% of all fruit decay is caused by species of *Penicillium* and 36% of all vegetable decay by a soft rot bacteria (Wiant and Bratley, 1948).

There have been many developments in the last two decades in the use of chemicals, antibiotics, and fungicides, concomitant with packaging materials and/or senescence inhibitor(s) to control microbial growth and at the same time retard the metabolism in fresh produce to extend the shelf life of the product. Thus in an increased number of products the beneficial effects of chemical-packaging have been demonstrated as a means of adding a safety factor against spoilage and of extending the quality for additional storage.

Fresh fruits and vegetables have short storage lives. They consist of senescencing and metabolizing tissues. This means research on radiation, chemical treatments, and packaging of fruits and vegetables to increase their storage life. We are concerned with inactivation of microorganisms and inhibition of physiological processes of the fresh produce, particularly biochemical changes such as respiration, phosphorylation, membrane permeability, and shift in protein synthesis.

As far as fruits and vegetables are concerned, the site of infection is primarily on the surface of the product. Surface pasteurization of the products without affecting quality of the product is essential. However, if the product is injured, the cell sap oozes out. Consequently there is a discoloration of the produce because of oxidation of substrates such as phenols and/or desiccation of the product; this also provides good nutrient media for microorganisms to grow and multiply.

Major sources of pathogen inoculum are picking and storage boxes, the atmosphere in packing houses and storage rooms, cleaning vats and hydrocoolers, and commodity-processing equipment. Adams and Tamuro (1959) isolated common decay fungi from apple- and peach-picking boxes and found that postharvest decay of these fruits could be substantially reduced by spraying the boxes with protectant fungicides.

It is standard commercial practice to fumigate stored grapes at 10-day intervals with 0.25% sulfur dioxide to prevent the spread of *Botrytis* (Harvey, 1956; Harvey and Pentzer, 1960). Sodium bisulfite is added to packaged grapes to prevent growth and spread of *Botrytis* during storage and long distance shipment (Pentzer, 1939). Dibromotetrachloroethane has shown considerable promise for this purpose (Chiarappa et al., 1962; Nelson et al., 1963). A substantial reduction of mycoflora was achieved by sprinkling grapes with $K_2S_2O_5$ followed by SO_2 fumigation, during a 4-month storage period at $0°C$ (Novobranova, 1970). Sulfur dust application on peaches after harvest reduced infection by *M. fructicola* (Cardinell and Barr, 1952). Thiourea and thioacetamide as 5% aqueous solutions have provided highly effective control of *Penicillium* decay as well as of *Diplodia* and *Phomopsis* stem end rots on oranges (Childs and Siegler, 1946; Hopkins and Loucks, 1946). Szkolnik and Hamilton (1957) reported that brown rot of peaches was reduced by dipping the fruits in a 1% suspension of the zinc salt of pyridine-N-oxide-2-thione. Development of *Penicillium expansum* was significantly reduced in stored pears by a 20-sec dip in a 500 ppm solution of thiabendazole (Spalding, 1970).

The use of biphenyl to control citrus fruit decay during shipment has been a major factor in the development of world trade in citrus fruits. Biphenyl vapors are strongly inhibitory to mycelial growth of *P. digitatum*, *P. italicum*, *Diplodia natalensis*, *Phomopsis citri*, *B. cinerea*, *Aspergillus*, *Monilinia*, *Rhizopus* and other plant pathogenic fungi (Ramsey et al., 1944; Heiberg and Ramsey, 1946). Tomkins (1936) observed that biphenyl vapors inhibited sporulation of *P. digitatum* on the surface of decayed oranges. This ability of biphenyl to control sporulation of *Penicillium,* and thus the spoilage of adjacent fruit by fungus spores, has made it possible to successfully market unwrapped citrus fruits from California (Eckert, 1959; Rygg et al., 1961). Farkas (1938, 1939) showed that untreated oranges developed 6 or 7 times more *Penicillium* decay than those wrapped with biphenyl-impregnated tissue paper. The efficacy of biphenyl for control of *Penicillium* has been confirmed by large-scale commercial test shipments (Roistacher et al., 1960; Rygg et al., 1961). Biphenyl also reduced decay of citrus fruits caused by the stem end rot fungi, *D. natalensis* and *P. citri* (Godfrey and Ryall, 1948; Littauer, 1956; Winston, 1948). Biphenyl has been tested for the control of *Botrytis* on grapes (Tomkins, 1936), *Rhizopus* on peaches (Haller, 1952), and *Fusarium* on potatoes (McKee and Boyd, 1962) with favorable results.

Sharma (1936) showed that a dissociated form of o-phenyl-phenol was nonphytotoxic and described a method for treating fruit with a solution of sodium o-phenylphenate containing excess sodium hydroxide to suppress hydrolysis of the o-phenylphenate ion. Treating the picking boxes with a solution of 8-hydroxyquinoline reduced the decay of strawberries (Jarvis, 1960).

Sodium salicylanilide is highly effective for control of *Nigrospora sphaerica* on bananas and has been used on a commercial scale for this purpose in Australia (Simmonds, 1949). It has been found in Australia that sodium salicylanilide is more effective than sodium o-phenylphenate for control of *Phomopsis* stem end rot of oranges (Hall and Long, 1950).

Bottini (1927) first reported that ammonia gas was effective in preventing decay of citrus fruits. Constant low concentration of ammonia gas (50 to 200 ppm) effectively controlled *Penicillium* decay and yet was not injurious to citrus fruit (Eckert et al., 1963; Roistacher et al., 1955, 1957). Ammonia fumigation has been suggested as a way to retard *Rhizopus* rot in peaches (Eaks et al., 1958). Solutions of 2-aminobutane salts are effective for control of *Penicillium* sp. on citrus fruits and apples, *M. fructicola* on peaches (Eckert and Kolbezen, 1964; Smoot and Melvin, 1964; McCornack and Hopkins, 1965; MacLean and Dewey, 1964; Jarrett and Gathercole, 1964; Salem et al., 1970).

Ozone has been tested as a fumigant for control of postharvest decay of fruits and vegetables. Concentrations of ozone tolerated by treated commodities may not control decay of inoculated fruit (Barger et al., 1948; Hopkins and Loucks, 1948), but may prevent the growth of surface molds, reduce the spores in the storage rooms, etc. (Schomer and McCollock, 1948). Fumigation with ozone has been reported to control *Botrytis* (Magie, 1961).

Nitrogen trichloride has been extensively used as a fumigant to control decay of citrus fruits (Littauer, 1956), melon (Barger et al., 1948), and certain vegetables (Pryor, 1959). Captan (N-trichloromethylmercapto - 4 - cyclohexene - 1,2-dicarboximide) has been evaluated as a postharvest treatment for control of decay during storage and marketing of strawberries (Becker et al., 1958; DiMarco and Davis, 1957a; Salunkhe et al., 1962), peaches (DiMarco and Davis, 1957b), cherries (Pierson, 1958), and pears (Pierson, 1960). Dipping the fruit in an aqueous suspension of 1,000 to 2,000 ppm DCNA (2,6-dichloro-4-nitroaniline) controlled *Rhizopus* on sweet cherries (Ogawa et al., 1961) and on peaches (Cappellini and Stretch, 1962). Domenico et al. (1972) found that Captan, Dithane, o-phenylphenate, and thiran have the greatest potential as fungicide treatments on tomatoes. The best method of application is to apply the fungicides in a wax solution. Benomyl is also good in preventing growth of most damaging fungi of tomatoes. Rahman et al. (1972) indicated that tomatoes treated with ClO_2 and then stored under low oxygen at $10°$ to $12°C$ exhibited the highest edible yield at the end of 6 weeks of storage. Captan was effective in controlling growth and stem scar of cantaloupes (Stewart, 1973).

Bacterial soft rot of packaged spinach could be delayed 2 to 3 days at $21°C$ by dipping in 1,000 ppm streptomycin (Smith, 1955). Even lower concentrations retarded the development of this disease on lettuce (Cox, 1955). Nystatin and primarcin have shown promise for control of *Monilinia* and *Rhizopus* on peaches (DiMarco and Davis, 1957b), strawberry rot (Salunkhe et al., 1962; Ayres and Denisen, 1958), and *Gloeosporium* on bananas (Meredith, 1960).

An analogue of vitamin K, vitamin K_5 (2-methyl-4-amino-1-naphthol hydrochloride), exhibits marked inhibitory activity toward a number of microorganisms, including common Brewer's yeast, Gram-positive, and Gram-negative microorganisms. Active concentrations of vitamin K_5 range from 0.05 to 0.005%. Vitamin K_5 prolongs the shelf life of a variety of foods and adds nutritive value to the food. Vitamin K_5 is a preservative with many merits for use in the fruit juice and other industries (Borgstrom, 1968).

Salunkhe and Norton (1960) noted that antibiotics such as Candicidin, Myprozine, and Amphotericin B inhibited fungal growth. However, Neomycin sulfate and D.P.S.-25 did not control the fungal growth. In addition to Captan, Dowicide A and DHA-S showed significant fungicidal effects on strawberries and cherries and maintained the natural color of the strawberries. A semipermeable polyethylene packaging material controlled secondary infection of the strawberries and cherries and modified the atmosphere around the fruits. This may have retarded respiration and transpiration which would increase the shelf life and help retain quality.

Do et al. (1966) conducted extensive research on the effects of hydrocooling 14 antifungal chemicals in vivo and in vitro and 5 packaging films on the refrigerated life and quality of sweet cherries. The hydrocooling along with antibiotics and packaging films revealed that fresh "Lambert" cherries were stored successfully for 30 days. In dip or spray treatments with Amphotericin B, Candicidin, Difolatan, Filipin, and Mycostatin over 90% of the treated fruits were marketable after 30 days' storage, whereas untreated controls (60% of the fruits) were unfit for consumption after 15 days of storage at 0 to 1°C and 90% R.H. Post et al. (1968) reported that phosphate compounds such as sodium hexametaphosphate, sodium tripolyphosphate, sodium tetraphosphate, and tetrasodium pyrophosphate significantly controlled the microbial growth on fresh cherries. In the in vitro studies sodium tetraphosphate appeared to be the most effective compound in preserving cherries and also had the greatest antimycotic effects against the most common fungal spoilers, *Penicillium expansum, Rhizopus nigricans,* and *Botrytis* sp.

Dehydroacetic acid has been used in recent years to reduce postharvest decay of fruits and vegetables (Borgstrom, 1968). It has proved to be a promising antimicrobial agent. The preservative action varies slightly with pH, has applications with many perishable foods, imparts no objectionable flavors at low concentrations, and is effective. Dehydroacetic acid is permitted for use as a pesticide only in special instances on strawberries in the U.S. (Desrosier, 1970).

CONTROLLED ATMOSPHERE (CA) STORAGE

Controlled atmosphere (CA) storage could be the most important innovation in fruit and vegetable storage since the introduction of mechanical refrigeration. This method, if combined with refrigeration, markedly retards respiratory activity and may delay softening, yellowing, quality changes, and other breakdown processes by maintaining an atmosphere with more CO_2 and less O_2 than in normal air. The long, involved, and unique history of CA storage illustrates the successful transfer of a scientific concept from laboratory to large-scale commercial use (Pantastico et al., 1970). Today over half the total holding of apples in the U.S. and citrus fruits in Israel are stored in CA.

Terminology

Technically, CA implies addition or removal of gases to reduce an atmospheric composition substantially different from that of normal air. Thus, CO_2, O_2, CO, ethylene, acetylene, or N_2 may be manipulated to attain various gas combinations. In common usage CA refers to increased CO_2, decreased O_2, and high N_2 levels as compared with normal atmosphere.

"Modified atmosphere" (MA) is often used interchangeably with CA. While MA storage, e.g., packaging in film bags, also requires a decrease in O_2 and an increase in CO_2 or N_2, there is no attempt to control the gases at specific concentrations. MA and CA differ only in degree of control and methods of maintaining that control.

"Gas storage" is an inadequate term to distinguish CA or MA from normal air storage. All three terms connote gas compositions. However, when only one gas is used, it is appropriate to indicate the gas involved. For instance, if 100% N_2 is substituted for normal air, the method may be termed "nitrogen storage." "Partial vacuum," "subatmospheric pressure," and "hypobaric" storage are synonymous terms. Hypobaric storage is gaining popular usage. It is a type of CA storage with emphasis on reducing the pressure exerted on the storage material. This method not only reduces O_2 concentration but also increases the diffusion of ethylene by evacuation from the tissues of the fruit, consequently extending the storage life (Wu et al., 1972).

Production of Controlled Atmosphere (CA)

Artificial atmospheres have been maintained in storage rooms in several ways. In the first method tried, fruit was merely placed in a sealed, airtight room. The respiration of the fruits increased the CO_2 and decreased the O_2 in the room. Proper levels were maintained by CO_2 scrubbers, usually NaOH. Smock and Yatsu (1960) found that water could be used for scrubbing CO_2 from CA storage when the concentration became too high. This method was less expensive and less noxious than NaOH.

With the increased interest in CA storage, large corporations have been experimenting with externally generated CA (Lannert, 1964). Currently Tectrol (Total Environment Control) is shipping fresh produce such as lettuce in refrigerated vans under CA. The system involves the introduction of properly proportioned blend of gases for specific fruits and vegetables into refrigerated containers as soon after picking as possible. It has been described in a patent assigned to Whirlpool Corporation, St. Joseph, Michigan. The gas composition consists of 1 to 5% carbon monoxide, 1 to 10% oxygen and 1 to 5% carbon dioxide with the remainder being nitrogen. Excess CO_2 is scrubbed by lime. The maintenance of the low oxygen atmosphere depends upon a dynamic relationship between lettuce respiration which consumes oxygen and a low leakage of air into the refrigerated container to replace it.

The Tectrol generator can produce sufficient CA to store from 12,000 to 18,000 bushels of fruits. The atmosphere can be reduced to the required level in about three days or one sixth of the time required by conventional CA. This results in less deterioration of the fruits. A storage chamber can be opened for removal of part of the stored fruits and quickly brought back to the proper atmosphere.

Another commercial CA system is Oxytrol. The Oxytrol system is a complete self-contained atmosphere control system designed to be used as an adjunct to normal refrigeration equipment in conventional transport vehicles. Liquid nitrogen supplied from a portable container is used to reduce the oxygen level. The van is insulated and equipped with a refrigeration unit to control temperatures. Automatic operation of the N_2 flow is provided by special controls in the system. CO_2 released during respiration is maintained at a low level by means of hydrated lime (scrubber). The container is somewhat tightly sealed to prevent excessive leakage of O_2.

Smock and Blanpied (1966) compared the effects of internally generated conventional CA and externally generated Tectrol CA. The conventional CA used lime [$Ca(OH)_2$] and water scrubbers to maintain the CO_2 level. They reported no difference in keeping quality. The longer periods of time required to establish the desired atmosphere as in conventional CA, however, made the atmospheres produced rapidly, as by the Tectrol equipment, superior.

The cost for materials and operation of a sophisticated externally generated CA system like the Tectrol generator may be rather costly for those developing countries in Southeast Asia which need this type of facility to extend the shelf life of fruits and vegetables they produce. Controlled atmosphere storage, however, could be produced by utilizing CO_2 evolved from respiration of the stored fruits and vegetables by controlled diffusion of the storage atmosphere over unopened bags of hydrated lime placed in a separate gas-tight container connected to the store by diffusion valves (Eaves, 1958, 1962). It is possible to regulate the concentration of CO_2 and O_2 by controlling the rate of diffusion with gate valves. When lime reacts with CO_2, water is released and heat is absorbed, and thus storage humidity is maintained. The maintenance costs are considerably lower.

Metabolic Effects of CA

Respiration

Many of the early investigators of CA storage recognized that it caused changes in the basic metabolic processes of the stored fruits.

Respiration appears to be one of the metabolic processes influenced by CA storage. Blackman and Parija (1928) postulated the following hypothesis. CA can influence respiration at three levels: (a) aerobic, (b) anaerobic, and (c) a combination of the two. Aerobic respiration occurs when the O_2 supply is normal and results in the liberation of CO_2 and water. Anaerobic respiration taking place in atmospheres completely devoid of O_2 produces CO_2 and ethyl alcohol by fermentation. When O_2 is low both processes function in proportion to the relative concentration of O_2. Aerobic respiration and sugar utilization increase as O_2 increases. A point may be reached at which minimal CO_2 is liberated from minimum respiration, and thereby minimum sugar utilization takes place. The metabolic processes of fruits are so complicated, however, that neither high CO_2 nor O_2 is completely satisfactory in changing the chain of reactions involved in respiration. This can be evidenced by the off-flavors, failure to soften properly, or even the complete breakdown of tissue before the normal ripening processes are completed (Claypool and Allen, 1947).

Kidd and West (1930) also recognized that CA had an effect on the respiration of stored fruit. They found that the average rate of loss of carbohydrates by respiration was between 1.2 and 1.4 times faster in normal air than in 10% O_2 and N_2, and between 1.35 and 1.55 times faster in the absence of CO_2 than in the presence of 10% CO_2.

Thus, fruits removed from conventionally refrigerated storage exhibited a higher rate of respiration than did CA-stored fruit (Littlefield, 1968). The rate of respiration of fruit removed from conventionally refrigerated storage increased rapidly for about eight days. The fruit from CA storage increased in respiration soon after removal from treatment; however, this increase was generally not as accentuated. Karnik (1970) and Karnik et al. (1970) similarly mentioned that sugar beets in CA storage had a lower rate of respiration than those stored in conventional refrigeration at 2°C. Littlefield (1968) found that the rate of CO_2 evolution in CA-stored "Lambert" sweet cherries was lower than that of conventionally stored fruit at 0°C. Controlled atmosphere (2.5% CO_2 and 2.5% O_2) alone and when combined with polyethylene packaging reduced the rate of respiration, while CA combined with Phaltan (1,000 ppm) enhanced the rate of respiration of the lettuce (Yang, 1971).

Mattus (1950) found that production of CO_2, ethylene, and other oxidizable volatiles was higher when "Bartlett" pears were stored in air than in CA. This is significant when one considers that the storage life of "Bartlett" pears terminates after their ethylene production peaks. Retarding the occurrence of this peak may extend storage life.

Hill (1913) observed that the rate of respiration of peaches did not return to normal after storage in CO_2 for a few days. McKenzie (1931) noted that as CO_2 was increased in storage, the rate of CO_2 evolution in lettuce decreased. Claypool and Allen (1948) reported that when O_2 levels were reduced below the atmospheric air levels during the transit periods, the respiration rates of apricots, grapes, peaches, pears, and plums were reduced. Lieberman and Hardenburg (1954) also noted a reduction in the rate of respiration when broccoli was stored at 24°C under various O_2 concentrations, especially as O_2 concentration was lowered from 20.9% to 0%.

Young et al. (1962) found that CO_2 delayed the onset of respiration rise in avocado and reduced the rate of O_2 uptake at climacteric peak. He speculated that reduced O_2 might delay the induction of the climacteric by decreasing the available ATP for synthesis. Increased CO_2 might delay the formation of a particular amino acid necessary for the synthesis of a specific enzyme, or it might delay the decomposition of an enzyme inhibitor. Also, an alternate metabolic pathway may be used.

Accumulation of Acids

Ranson (1953) observed no major changes in the acid contents of kalanchoe leaves (12 hr), carrot roots (3 days), and oat coleoptiles (1 day) after storage in atmospheres containing less than 10% CO_2 at 19°C; however, in atmospheres containing between 20 and 90% CO_2, succinate and aminobutyrate accumulated; while malate, aspartate, and alanine were depleted. Littlefield (1968) found that organic and amino acids generally reacted similarly in apples and pears and both types tended to accumulate in CA-stored fruits. Wankier (1970) and Wankier et al. (1970) noted that succinic acid accumulated in apricots and peaches when the concentration of CO_2 was above that of normal atmospheric air at 1°C, while maleic and citric acids progressively decreased during such storage. Littlefield (1968) and Singh et al. (1970) found that "Lambert" sweet cherries stored at a 10.5% CO_2 concentration had less tyrosine and more α-aminobutyric and maleic acids than those stored in conventional refrigeration at 1°C.

Ranson (1953) and Hulme (1956) found that succinate accumulated in plants stored in atmospheres containing more than 20% CO_2. Later, Ranson et al. (1957) pointed out that the accumulation of succinate was due to succinic oxidase sensitivity to higher levels of CO_2. Frenkel and Patterson (1969) found that the activity of succinic dehydrogenase was inhibited if the fruit was kept in CO_2 concentrations above 0.03%. Claypool et al. (1966) exposed apricots to atmospheres containing air, 2.5, 5, and 10% CO_2 and demonstrated less acid loss, as measured by pH and titratable acidity, at all CO_2 levels as compared to air storage. Thornton (1937) exposed apples to concentrations of CO_2 from 0 to 60% along with 20% O_2 and found (after 10 days at various temperatures) no changes in the ascorbic acid content.

Groeschel et al. (1966) found that the pH of CA-stored green beans increased in two weeks, while that of air-stored samples at 7°C remained fairly constant. The titratable and total acid anion contents increased during air storage at 7°C, but decreased slightly in CA. Burgheimer et al. (1967) found a marked increase in pH and a decrease in the titratable acidity of CA-stored spinach. Leberman et al. (1968) also found that the pH of broccoli increased progressively with increased concentration of CO_2 in the storage atmosphere.

The O_2 level did not affect the pH. An increase in pH was paralleled by a decrease in titratable acidity. Wankier et al. (1970) observed that the pH of stored apricots and peaches was not affected by CA storage at $1°C$. The pH and titratable acidity were inversely related and titratable acidity decreased with 15 to 60 days of storage. A marked increase in pH paralleled by a decrease in titratable acidity has also been noted in asparagus stored in CA (Wang et al., 1971).

Li (1963) cited as primary features of metabolic changes caused by CA that while metabolic pathways conform, specific differences occur. He stated that modifications of conventional atmosphere resulted in specific alterations in metabolism, and certain of these induced changes appeared to be reflected in improved quality of the fruit. He found that after seven months of storage there was 78% more malic acid in the CA-stored "Anjou" pears. Allen and Smock (1937) also found that CO_2 atmospheres seemed to check the decrease in acidity usually experienced in storage. He suggested that acid accumulations during CA storage might be due to lowered respiratory activity, increased CO_2 fixation, or the presence of a less active enzyme that converts malic acid to pyruvate or oxaloacetate. Goddard and Meeuse (1950) found that as CO_2 concentrations of the atmosphere increased, CO_2 fixation also increased. Allentoff et al. (1954) identified C^{14} fixation in malic acid and in the three key amino acids, glutamine, aspartic acid, and alanine.

Acetaldehyde

Thomas (1925) found that acetaldehyde was produced in large quantities in certain mixtures of CO_2 and O_2 with high concentrations of the substance often accompanied by browning of the cells. Trout (1930) found 2% O_2 to be in excess at $3°C$ which was still above the region of extinction by the anaerobic process. These tests were made by determining the amounts of alcohol and acetaldehyde present in the fruit tissue.

Sugars

Li (1963) investigated many aspects of the biochemical changes instigated by the use of CA. Using 2% O_2 and 3% CO_2 he found that starch disappeared from "Bartlett" and "Anjou" pears during the first month in storage. The sugar content showed no consistent differences except

that sucrose appeared to decrease more rapidly in CA, but not significantly. The CA-stored fruit had more sugar after ripening. Conventional storage produced relatively no changes in sugars.

Alcohol Soluble and Protein Nitrogen

Total alcohol soluble nitrogen, decreased at a slower rate in fruit in CA storage resulting in higher concentrations of alcohol-soluble nitrogen at the end of the storage period (Li, 1963). Protein N responded the same way; i.e., it decreased at a slower rate in the fruits stored in CA.

Pectins

Pectic substance changes were correlated with changes in the storage atmospheres (Smock and Allen, 1937). Soluble pectins increased in "Bartlett" and "Hardy" pears stored at $7°C$, but in "Yellow Newton" apples the increase of soluble pectins was retarded by CO_2 storage. In "Bartlett" pears a significant correlation was found between degree of maturity and content of soluble pectins (Gerhardt and Ezell, 1938). It was suggested that CO_2 affected protopectin hydrolysis. Gerhardt and Ezell (1938) showed that the soluble pectin content was much higher where higher temperatures or no CO_2 was involved. Wankier et al. (1970) also pointed out that although the rate of the pectin degradation was affected by both time and conditions of storage, the CA-stored apricots and peaches retained a higher concentration of total pectins than did conventional refrigerator-stored fruit.

Chlorophyll

McGill et al. (1966) observed that the effects on total chlorophyll content of spinach could be related to storage atmosphere, duration, or temperature. Groeschel et al. (1966) reported that the presence of CO_2 in the storage atmosphere was important in preventing chlorophyll degradation of green beans at $35°C$. Leberman et al. (1968) observed that chlorophyll retention was increased in broccoli by a progressive increase in CO_2 and decrease in O_2. Wankier et al. (1970) noted that higher concentrations of CO_2 (2.5 to 10%) slowed down the degreening process in apricots and peaches. It has also been demonstrated that CO_2 suppressed chlorophyll destruction in apricot and peach fruits at $1°C$. The rate of degradation of chlorophyll in asparagus was also delayed by CA storage (Wang et al., 1971). Wang (1971) and

Singh et al. (1972a) observed the inhibition of chlorophyll degradation in lettuce, *Lactuca sativa* L., Cultivar "Great Lakes," throughout a 75-day storage period. Storage conditions were a CA consisting of 2.5% CO_2 and 2.5% O_2 at 2°C with or without prestorage treatment with Phaltan (1,000 ppm) or polyethylene packaging.

Adverse and Toxic Effects of CA

Fulton (1907) was the first to note fruit damaged when a large amount of CO_2 was present in the storage room. Controlled atmosphere storage has shown great promise, but adverse effects have been observed. The biochemical constituents of fruit have since been shown to change in an unusual manner during CA storage. Injury to fruit tissue may be caused by an abnormality in metabolism induced by high CO_2 and low O_2 concentrations. Some of these disorders are in the form of browning of the fleshy mesocarp, tissue breakdown, and the accumulation of certain organic acids (Smock, 1958b; Hulme, 1958). Hulme (1956) stated that succinic acid, even in concentrations of less than 0.001 M, was toxic to fruit. Conducting experiments on apple peel, he demonstrated that when succinate was supplied to the respiring peel, the respiration fell to zero and the tissue became brown and "dead." Smock (1958a) stated that if there is a complete lack of O_2 for a period of several days, the fruits will suffocate and become alcoholic and badly off-flavored. Miller and Evans (1956), Miller et al. (1960), and Porter and Thorne (1955) indicated that an increased CO_2-bicarbonate mixture markedly affected the rates of several enzyme reactions.

Ranson et al. (1957) conducted experiments with ^{14}C-labeled substrates and followed the oxidation of these substrates in the presence of CO_2-bicarbonate mixtures. They observed that the production of malate from succinate oxidation was diminished by about 40% in 10% CO_2 and by more than 30% in 5% CO_2. Miller and Hsu (1965), using the indophenol reductase system, demonstrated a 17% inhibition with 5% CO_2, and 30% inhibition with 15% CO_2 treatment. Ranson et al. (1957) hypothesized a general sensitivity of all dehydrogenase to CO_2-bicarbonate mixtures but concluded that other experiments would need to be conducted with ^{14}C-labeled Krebs cycle intermediates before the theory could be proven.

Tissue browning is another common disorder of fruit. One main cause of such discoloration is biochemical changes in tannins. The tannins are a complex assemblage of polyphenols which often have no close structural relationship. Tannins can be distinguished from other phenols by their precipitating action on proteins; however, this does not apply to the caffetannins such as caffeic acid and chlorogenic acids (Russell, 1935).

Pridham (1956) stated that most phenolic acids are phytotoxic but are probably harmless to the plant because in the native state they are glycosides or esters. If during storage and senescence these compounds were hydrolyzed, toxic products could be accumulated in the tissues, possibly killing cells and serving as substrates for enzymatic browning reactions. Bate-Smith (1954) and Swain (1962) indicated that polyphenolic compounds are important fruit constituents related to color, taste, and enzymatic browning. Onslow (1920) showed the presence of phenolic compounds in peaches and other fruits and identified the catechol tannin as a substrate for the oxidizing enzyme in several fruits. Johnson et al. (1951) isolated the following tannins in peaches: D-catechin, phlabatannins, and chlorogenic acid. Bate-Smith (1954) stated that when both active polyphenoloxidase and phenolic substrate are present, the action of one upon the other produces a discoloration of the tissue, usually a browning. Harel et al. (1966), reviewing the work of other investigators, stated that the intensity of browning is determined by the amounts of phenolic substances present in the fruit. Walker (1964), comparing the intensity of darkening of different cultivars of apples treated with chlorogenic acids, came to a similar conclusion.

Reeve (1959) conducted studies on histological and histochemical changes in ripening peaches and demonstrated by the nitroso-reaction that the endocarp tannins increased with ripening. The endocarp tissue showed a much higher concentration of tannins than did the mesocarp tissue. He also indicated a characteristic localization of tannins in patches of enlarged mesocarp parenchyma cell. Bezlur and Joslyn (1954) found that prune mesocarp tissue near the endocarp was higher in enzyme activity, and that both phenolase and peroxidase enzymes were involved in tissue browning. Williams (1961) observed in apples that phenolic compounds were twice as high in the core area as the mesocarp tissue. El-Tabey and Cruess (1949) stated that both peroxidase and phenolase were active in apricot browning. Phenolase activity increased from zero at pH 2.2 to a maximum at pH 7.

Brown stain of lettuce is a physiological disorder associated with increased CO_2 in the range of 2.5 to 10% (Stewart et al., 1970). The symptoms include variable sized reddish-brown pitted spots along the midrib. The market quality of head lettuce has been evaluated after storage in low O_2 atmosphere which retarded russet spotting but increased brown stain (Stewart and Uota, 1971). Carbon monoxide (CO) in the presence of elevated CO_2 increased lettuce susceptibility to brown stain (Stewart and Uota, 1972). This was further substantiated by Kader et al. (1973). They found that brown stain on crisp head lettuce was increased by exposure to CO combined with elevated CO_2 (1.2 or 5%) regardless of the O_2 level (2.5, 10, or 21%). Carbon monoxide at 1.3 or 5% added to air without CO_2 did not induce brown stain.

Effects of CA on Delaying Senescence and Extending Shelf Life

The first scientific investigation on the effect of atmosphere on fruit ripening was conducted by Berard in 1821. According to Berard, the fruits placed in an atmosphere deprived of O_2 did not ripen. By placing an O_2 and a CO_2 absorbent (lime, $FeSO_4$, and water) and fruit in a jar, he observed that peaches, prunes, and apricots could be stored from 20 days to 1 month, and pears and apples about 3 months. The results may have been due to both low-O_2 and low-CO_2 levels which is now known as a low-O_2 atmosphere.

Brooks and Cooley (1917) indicated that in an atmosphere containing 5% CO_2, apples remained green, firm, crisp, and slightly alcoholic for 5 weeks. A more general study of the storage of apples by Magness and Diehl (1924) indicated a marked relationship existed between CO_2 concentration and the softening of the fruit. An atmosphere containing over 5% CO_2 resulted in a distinctly slower rate of softening than one of normal concentration.

Trout (1930) realized the advantages of using CA to extend shelf life by retarding the ripening process. Allen and Smock (1937) found that 10% CO_2 at $0°C$ kept apples in as good a condition as did air storage at $0°$ to $2°C$. Fisher (1942) tested several atmospheres with "Golden Delicious" apples and found 5% CO_2 and 2.5% O_2 main-

tained the fruit in good condition for several months. Huelin and Tindale (1947) discovered that although gas storage at some concentrations may cause scald, gas storage also controlled "Jonathan" spot and bitter pit. Allen and Claypool (1948) used 5 to 10% CO_2 and 2.5% O_2 at $0°C$ and found that pears remained in good condition and ripened satisfactorily after nearly 6 months in storage. This amounted to almost two times the usual storage period. After becoming fully ripe there was little difference in the time each remained marketable.

Plant activity can be slowed down by modified storage atmospheres. There are limitations, however, and these may vary with different fruits and with different cultivars of one fruit (Brooks, 1940). In CA storage using CO_2 gas, it is important to determine the tolerance of various plant organs to the artificially produced CO_2 atmosphere.

Watada et al. (1964) reported that lettuce exposed to an increased CO_2 atmosphere for 2 to 4 days appeared slightly better than that held in air. The cut ends of stem and ribs were white and free of blemishes and browning. But after removal from the CA and after several days in air at $10.5°C$, the symptoms of CO_2 injury were apparent, viz., variable-sized reddish spots along the midrib, yellow-brown slimy discoloration of the cut stem with rot in its center, desiccated leaves with yellow oily droplets of exudate, and general softening of the tissue.

In-transit tests of lettuce shipped in refrigerated trailers or mechanically refrigerated trailers with supplemental nitrogen were made by Stewart et al. (1966). They found that low-O_2 atmosphere reduced russet spotting but had no effect on decay. Physiological injury sometimes developed on the wrapper leaves of lettuce when the O_2 concentration was below 1% for more than 42 hr. Lipton (1967) also pointed out that russet spotting was reduced in low-O_2 atmospheres and butt discoloration was generally less intense in 0.5 and 1% O_2 than in air, regardless of temperature. Rahman et al. (1969) determined the effects of low-O_2 (Oxytrol-CA) on lettuce for 22 days at $1°$ to $2°C$. Stewart et al. (1970) observed that the incidence of pink rib was slightly lower in lettuce shipped in modified vehicles (5% O_2) than in lettuce in the normal atmosphere vehicles at $2.8°C$. Yang (1971) and Singh et al. (1972b) reported that lettuce heads of cultivar "Great

Lakes" could be stored for up to 75 days in a controlled atmosphere of 2.5% CO_2 and 2.5% O_2 at $1.7°C$ and 90-95% R.H. Prestorage treatments with microbe- or senescence-inhibiting chemicals (Captan, 1,000 ppm; Phaltan, 1,000 ppm; Mycostatin, 400 ppm; and N^6-benzyladenine, 20 ppm) had detrimental effects on lettuce in CA storage at $1.7°C$. Neither dry weight nor moisture content of the lettuce, however, was affected by the CA or the other treatments.

The effects of CO_2 on respiring plant organs were the subject of investigations of ways to retard respiration. As early as 1930, results of many short-period experiments were reported. With few exceptions, CA storage surpassed chemical or other treatments because it retarded respiration without causing any apparent injury to living plant organs. Fruits and vegetables can be held in storage for 3 to 7 days in concentrations of CO_2 varying from 6 to 83% without apparent injury, depending upon the variety tested. Tolerance for CO_2 was influenced by ripeness, firmness, and freshness of the plant organ (Thornton, 1930). Kidd et al. (1927) demonstrated that 10% CO_2 extended the storage life of apples 1.5 times beyond that of the control.

In the storage of Danish cabbage, Isenberg and Sayles (1969) found that the best combinations of gases ranged between 2.5 and 5% of O_2 and CO_2, with N_2 making up the remainder. These combinations reduced weight loss due to respiration and trimming loss. They also reported that with the atmospheres of low O_2 (2.5 to 5%) and high CO_2 (2.5 to 5%), cabbage heads retained their green color, succulence, and firmness.

Chawan and Pflug (1967) studied CA storage of "Downing Yellow Globe" and "Abundance" cultivars of onions with different CO_2 and O_2 levels at several temperatures. They found that best results were obtained when onions were stored at $4.4°C$ in either 10% CO_2 and 3% O_2 or 5% CO_2 and 5% O_2. The CA-stored onions appeared to be superior to the onions stored in air at $5.5°C$.

Littlefield (1968) reported that atmospheres consisting of 1.5% CO_2 and 2.5% O_2 for apples and pears and 10.5% CO_2 and 2.5% O_2 for cherries inhibited the metabolic system and thus extended shelf life. Anderson et al. (1969) generally found less decay occurred in peaches and nectarines stored in 5% CO_2 than in those at 0%

and at $0°C$. Wankier et al. (1970) and Wankier (1970) noted that firmness, total pectins, titratable acidity, total sugar, and tannins changed at slower rates in CA-stored apricots and peaches than in those stored in the conventional refrigerated room.

Salunkhe and Wu (1973a) reported that low oxygen atmospheres at $12.8°C$ inhibited the ripening and thus increased the storage life of tomatoes. At 10% O_2 and 90% N_2, the storage life was 62 days, while at 3% O_2 and 97% N_2, it was 76 days. The maximum storage life of tomato fruits was 87 days at 1% O_2 and 99% N_2. Low oxygen atmospheres inhibited degradation of chlorophyll and starch, and syntheses of lycopene, β-carotene, and soluble sugars of tomato fruit.

Subatmospheric Pressure Storage

It has been known for many years that the ripening of fruits is associated with their production of ethylene in the presence of an adequate oxygen supply. When either oxygen or ethylene is reduced, ripening is usually delayed, and the marketable life of the produce is subsequently extended. Apples are now stored for extended periods in controlled or modified atmospheres. Subatmospheric pressure (hypobaric) storage is a type of controlled atmosphere storage with emphasis on reducing the pressure exerted on the storage material. Workman et al. (1957) noted a trivial reduction in the respiration of tomato fruits when the surrounding storage pressure was increased to 120 cm of water. Hummel and Stoddard (1957) reported increases of 20 to 92% in the life of several kinds of produce stored in a home-type refrigerator with a compartment maintained at a pressure between 658 and 709 mm of Hg. Burg and Burg (1966) obtained similar success with additional kinds of fruits and vegetables stored at pressures from 125 to 360 mm of Hg and in air. They suggested that intercellular ethylene in fruits, prior to respiratory climacteric, rises to levels that stimulate ripening; and if this ethylene is removed, ripening is delayed. Tolle (1969), working with mature green tomatoes, found that the lowest pressure tested (180 to 190 mm of Hg) retarded fruit ripening the most when the oxygen partial pressure was also reduced. When oxygen pressures were held constant, different levels of total hypobaric pressure caused little if any difference in retardation. The retardation of ripening and color development of mature green tomatoes may be more directly related to oxygen levels than to levels of total pressure and reductions of ethylene.

In an extensive study, Wu et al. (1972) found that the ripening of "green-wrap" tomato fruits was retarded and the storage life extended by subatmospheric pressures. The ripening coefficients of treated tomatoes are significantly (1% level) different from those of controls. Under normal conditions (control), "green-wrap" tomatoes ripened in 35 days at 13°C; tomatoes stored at 471 mm Hg ripened in 65 days, and those stored at 278 mm Hg ripened in 87 days. Tomato fruits never ripened, but did deteriorate if stored at 102 mm Hg beyond 100 days. Tomato fruits previously stored at 102 mm Hg for 100 days ripened normally in 7 days when transferred to 646 mm Hg (atmospheric pressure at Utah State University, Logan, Utah) at 13°C and 90 to 95% R.H. The loss of chlorophyll from the fruit was delayed by storage at subatmospheric pressures. The lower the pressure the greater the delay.

Loss of chlorophyll proceeded normally when fruits stored at 102 mm Hg were placed at 646 mm Hg. Synthesis of lycopene and β-carotene of tomato fruits was also inhibited by subatmospheric pressures. The lower the pressure the longer the inhibition, especially of lycopene. The formation of lycopene is negligible in early stages of storage, but becomes significant later. With storage at 102 mm Hg, lycopene formation was completely inhibited for 100 days. After transfer to 646 mm Hg, lycopene formation was initiated. The formation of β-carotene is less inhibited by subatmospheric pressure than that of lycopene. Subatmospheric pressure storage inhibited starch degradation and sugar formation. Under normal conditions starch decreased and sugar increased during the ripening of tomato fruits. The inhibition of this process was more distinct in the early stages than in later stages. At 102 mm Hg, this inhibition lasted 90 days. After 100 days' storage at 102 mm Hg and then 7 days at 646 mm Hg the sugar content of the red ripe fruit was significantly (at the 5% level) less than that of fruit stored at 646 mm Hg for 35 days. Wu et al. (1972) attributed the inverse relation between the amount of volatiles and pressure to a continuous evacuation of flavors in the gaseous state. Compared to the controls, tomatoes under high vacuum for 100 days and subsequently ripened at

atmospheric pressure had the least amount of aroma components. This may be due to a removal of flavor compounds and their precursors under vacuum. In tomato fruits, decreases in chlorophyll and starch and increases in lycopene, β-carotene, sugar, and flavor accompanied the ripening process. The inhibition of these changes under subatmospheric pressure can be attributed to the inhibition of the ripening process.

Salunkhe and Wu (1973b) noted that subatmospheric pressure storage significantly delayed softening and extended the storage life of apricots. The storage life of the control apricots was 53 days, while that of apricots stored at 102 mm Hg was 90 days. Subatmospheric pressure delayed formation of carotenoids beyond initial values. Formation of carotenoids was delayed for 60 days in apricots stored at 102 mm Hg. There was no significant difference in the total carotenoid contents of treated and control apricots at the end of the storage. Subatmospheric pressure slowed losses of sugars and acid; the lower the pressure, the slower the losses. The sugars and acidity of the treated apricots were almost the same as those of the control apricots at the end of storage. Softening of firm mature peaches was also significantly delayed by subatmospheric storage. The storage life at 102 mm Hg was 93 days, whereas that of control peaches was 66 days. In the control peaches, total carotenoids increased steadily during ripening. At 102 mm Hg, carotenoid formation was completely inhibited for 45 days. At the end of storage, control and treated fruits contained equal amounts of carotenoids. During ripening of the control peaches, sugar content increased somewhat, then decreased. The rate of such changes was significantly affected by the subatmospheric pressure storage, but at the end of the storage, there was no significant difference in the sugar contents of treated and control peaches. Titratable acidity of the peaches decreased during storage and these decreases were delayed by subatmospheric pressure. However, there was no significant difference in final titratable acidity between treated and control peaches.

Storage life of control cherries was 60 days, whereas that of the 102 mm Hg-treated cherries was 93 days. The pedicels of the treated fruits were green after 60 days of storage, whereas those of the control fruits were brown and moldy. Anthocyanins did not change even in conventional or subatmospheric pressure storage. As with apricots and peaches, subatmospheric pressure delayed sugar losses in sweet cherries.

Subatmospheric pressures significantly softened and extended the storage life of pears (Salunkhe and Wu, 1973b). Pears were stored for 3.5 months under normal refrigeration. At 461 mm Hg, they were stored up to 5 months; at 278 mm Hg, 7 months; and at 102 mm Hg, 8 months. The decrease of firmness was especially delayed at 102 mm Hg. The green color of the pears was retained fairly well up to 5 months at 102 mm Hg. Degradation of chlorophyll was delayed by subatmospheric pressure. The lower the pressure the longer the chlorophyll was retained. Subatmospheric pressure also delayed losses of sugar. At the end of the storage, the sugar contents of pears from subatmospheric pressure were significantly less than those of the control fruits. The 278 mm Hg treatment extended the storage life of "Red King" apples for 3.5 months over that of the controls. With "Golden Delicious" apples, it extended the storage life for 2.5 months. There were no visual changes in the apples during storage, but the subatmospheric pressure storage delayed losses of sugars in both cultivars. After storage, however, treated apples contained significantly less sugars than did the controls. Decreases in titratable acidity of apples were also delayed by subatmospheric pressure storage.

Subatmospheric storage of potatoes can eliminate greening under certain conditions (Jadhav et al., 1973). Chlorophyll synthesis in the tubers stored at 126 mm Hg was completely inhibited, while 253, 380, 507, and 633 mm Hg pressure treatments were ineffective in controlling chlorophyll formation. There were no differences in the solanine levels of the tubers subjected to storage pressure treatments and the controls.

As to the effect of subatmospheric pressure on fungi associated with deterioration of fruits and vegetables during storage, Wu and Salunkhe (1972c) observed that subatmospheric pressure of 471 mm Hg has little or no effect on the growth and sporulation of the fungi. The growth and sporulation of these fungi were significantly suppressed at 278 mm Hg. The percent mycelial coverage of these fungi was less than that on the controls at this pressure. The appearance of growth was delayed and formation of spores was inhibited. Subatmospheric pressure of 102 mm Hg retarded the growth and sporulation of fungi. The lower the pressure applied, the more the inhibition

of growth and sporulation of fungi. For *Penicillium expansum,* percent mycelial coverage on petri dishes was 80 when stored at 102 mm Hg. The time for appearance of growth was almost twice as long as that of the control. At the end of incubation, control *Penicillium expansum* showed uniform dark blue spores, whereas treated *Penicillium expansum* showed only scattered light blue spores indicating slower growth of the fungus. The growth of *Rhizopus nigricans* was inhibited by 102 mm Hg. Mycelial coverage was 75% of that of the controls and the time for appearance of growth was two and a half times as long as that of the controls. Spore formation was also retarded. Five days after incubation most of the sporangia in the treated fungi were still white, whereas those of the control fungi were already mature and black. The growth and spore formation of *Aspergillus niger, Botrytis alli,* and *Alternaria* sp. were also retarded by 102 mm Hg. The time for appearance of growth was increased at this pressure. A controlled atmosphere of 2.7% oxygen and 97.3% nitrogen (646 mm Hg) had less effect on the growth and sporulation of the fungi than did subatmospheric pressures of 278 mm Hg and 102 mm Hg. It is apparent that with proper combinations of low temperatures, relative humidity, and correct gas composition, subatmospheric pressure should effectively control fungi and thus extend the storage life of perishable fruits and vegetables.

Modified Atmosphere Storage and Packaging

In a test of packaging treatments, Salunkhe and Norton (1960) and Salunkhe et al. (1962) found that polyethylene bags retarded respiration and transpiration and perhaps helped increase shelf life and retain quality of strawberries and cherries. Ayres and Denisen (1958) reported that the use of plastic containers generally resulted in less spoilage of strawberries and raspberries than when the fruit was stored in wooden boxes, especially if the wooden boxes had been used previously. They found cellulose acetate was the most promising film for wrapping strawberries and raspberries. Color and firmness of cranberries were better after 30 days or longer storage when the fruit was packaged in Mylar, Saran, or polyethylene. Hardenburg et al. (1958) stated that cherries in polyethylene may be held in cold storage at $-0.5°C$ for as long as 2 weeks following harvest and still be shipped by freight to eastern markets in acceptable condition.

The beneficial effects of modified atmosphere storage of "Nubiana" plums were reported by Couey (1965). The plums were stored in 1.5 mil polyethylene liners which provided modified atmospheres that averaged 7.8% CO_2 and 11.0% O_2. Decay, softening, and loss of soluble solids were reduced during storage period up to 10 weeks when compared with fruit in vented liners. "Eldorado" plums also stored well for 6 weeks in the modified atmosphere of 6.6% CO_2 and 7.2% O_2 provided by polyethylene liners (Couey, 1960).

Patil et al. (1971b) exposed "Russet Burbank" potato tubers to a light intensity of 100 fc for 5 days at 20.1°C after irradiation with gamma rays in CO_2-enriched (15%) clear polyethylene packaging. The dose of 10 krad of gamma radiation inhibited 80% of chlorophyll synthesis, while the polyethylene packaging with CO_2 inhibited only 33%. The irradiation and the CO_2 environment, whether alone or in combination, did not affect solanine synthesis.

IONIZING RADIATIONS

Atomic energy presents a multitude of possibilities in food preservation. One advantage of using ionizing radiation for fruits and vegetables is that it does not appreciably raise the temperature of the produce during the treatment. Thus, perishable fruits and vegetables can be preserved in a fresh state. Investigations on the influence of ionizing irradiation on potatoes by Sparrow and Christensen (1954) created some interest in the utilization of radiation to increase the storage life of potatoes and other root and bulb crops. Their observations on the inhibition of sprouting were confirmed by numerous workers (Mikaelsen and Roer, 1956; Sawyer and Dallyn, 1956; Brownell et al., 1957; Burton and Hannan, 1957; Schreiber and Highlands, 1958; Cloutier et al., 1959; Duncan et al., 1959; Madsen et al., 1959; Workman et al., 1960; Teixeira and Baptista, 1968). Beta and gamma radiations with doses of 1 to 5×10^5 rad range (radurization) have been shown to retard the growth of spoilage organisms such as bacteria and fungi for appreciable periods of time (Johnson, 1932; Hannan, 1954; Beraha et al., 1957; Salunkhe et al., 1959a; Beraha et al., 1960; Salunkhe and Simons, 1960; Cooper and Salunkhe, 1963).

Licciardello et al. (1959) showed that the products sterilized with irradiation when stored at 10° and 19°C were acceptable after 10 months' storage. Deterioration in quality occurred within one month when samples were stored at 51.5°C. Radiation-induced flavor diminished progressively as the storage duration advanced (Salunkhe et al., 1959c). Irradiation of potatoes (Pederson, 1956), onions (Sawyer and Dallyn, 1956), carrots (Madsen et al., 1959), and many other fruits and vegetables (Salunkhe et al., 1959b; Salunkhe and Simon, 1960) produced no detrimental effects on flavor when irradiated at the low doses.

Salunkhe et al. (1959a) found that sweet corn irradiated with husks was scored higher in taste preference than was sweet corn without husks, regardless of radiation dosage, storage duration, and temperature during radiation. Tomatoes and strawberries are too sensitive for gamma radiation. Ripe "Bartlett" pears could stand only small doses of radiation, while mature green pears could stand a moderate (up to 2.5×10^5 rad) amount without adverse effects of radiation. An intermediate rate $(0.465 \times 10^5$ rad/hour) has good effects on quality of asparagus and cabbage. Taste preference of irradiated asparagus, string beans, and cabbage was improved after 10 days' storage at 10°C. Mold growth was suppressed by gamma radiation on many products.

Ripening processes of green tomatoes were slowed by radiation. The threshold dose was in the vicinity of 1.86×10^5 rad. The red color of tomatoes and of a sour-cherry drink faded at doses of 5×10^5 to 1×10^6 rad. Anthocyanin and lycopene pigments were much more susceptible to the degradation by radiation than were the carotenoid and chlorophyll pigments. Radiation at 0.93×10^6 rad/hr tenderized asparagus much more than did the low $(0.093 \times 10^6$ rad) and intermediate rates $(0.465 \times 10^6$ rad). The ascorbic acid content of strawberries was reduced more drastically by the fast than by the slow rate at the low dosages.

Complex components of food such as proteins, fats, carbohydrates — starch, sugar, inulin, cellulose, and pectins — were degraded to their simpler components (Boyle et al., 1957; Brownell et al., 1957; Madsen et al., 1959; Salunkhe, 1957, 1959; Salunkhe et al., 1959a). This degradation of some of these compounds has been attributed to the softening of foods. However, when the products were irradiated at 2 to 3×10^5 rad, no unacceptable softening was noticed.

Salunkhe (1957) noticed the destructive effect of ionization radiations upon the green color of vegetables. A linear decrease in the chlorophyll content of green beans and broccoli resulted from increases in gamma radiation dose over the range of 0.49 to 9.29 Mrad (Wishnetsky et al., 1959). Lukton and MacKinney (1956) found the destruction of carotene and xanthophyll upon irradiation ranged from 5 to 95% in broccoli, 3 to 20% in sweet potatoes, and 0 to 5% in carrots.

Markakis et al. (1959) reported that 0.365 Mrad of 2,000,000 electric volt cathode rays destroyed 55% of the anthocyanins of strawberries and dehydration of the juice increased the radioresistance of the pigment. Potatoes (Madsen et al., 1959), lima beans (Salunkhe, 1957), and Jerusalem artichoke tubers (Salunkhe, 1959; Salunkhe et al., 1959b) became darker and progressively more brown as the radiation dose increased beyond 0.93×10^6 rad.

Panalaskas and Pellefier (1960) found that specified levels of gamma radiation did not cause consistent variations in the ascorbic acid contents of potatoes. Mikaelsen and Roer (1956) reported that the vitamin C content decreased in both irradiated and nonirradiated potatoes during the first 7 months of storage at 5°C but was restored after this period, and that ascorbic acid levels of the radiated samples were higher than those of the control. Kung et al. (1953) reported that vitamins A and C and tocopherols were sensitive to gamma radiation at sterilization doses. Vitamin analyses for thiamine, riboflavin, carotene, and pyridoxine indicated that the amounts of these vitamins destroyed by the sterilizing dose of cathode rays and X-rays were not as drastic as ascorbic acid.

In short, low doses of radiations have promising application in extending the storage life of fresh fruits and vegetables depending upon the crop, cultivar, maturity, and preirradiation antifungal treatment and packaging. For many fruits and vegetables, doses below 10 to 15×10^3 rad can reduce sprouting losses; those below 2×10^5 rad can effectively curb mold growth and delay ripening processes without causing much loss of quality and nutritive value.

OTHER IMPORTANT AND ASSOCIATED CONSIDERATIONS TO CONTROL RIPENING, SENESCENCE, AND MICROBIAL GROWTH

Temperature

Metabolic breakdown and fungal deterioration are two important factors in the spoilage of fruits and vegetables. Precooling — the prompt application of refrigeration prior to normal storage or transportation — is one key to successful storage. Precooling by hydrocooling removes the field heat of fresh fruits right after the harvest, and slows the metabolic breakdown as well as the activities of microflora, thus extending the storage life of the fruit. McClure (1958) reported that hydrocooling peaches with the addition of Clorox or Dowicide A to the hydrocooling water significantly reduced the decay caused by the brown rot or the *Rhizopus* rot fungus. Salunkhe et al. (1962) reported that hydrocooled peaches lasted longer than nontreated peaches. Do et al. (1966) found that hydrocooled cherries retained their natural bright red color fairly well for up to 30 days' storage.

Heat treatment appears to be the best and most convenient means of achieving postharvest disease control. Hot water has usually been used as the heat transfer medium for this purpose owing to its availability, heat capacity, and lack of residue problem. Fawcett (1922) was the first to point out the advantages of heated water for the therapy of a postharvest disease that was beyond control by fungicide treatment. He reported that infection of *Phytophthora citrophthora* could be eradicated from lemons after 30 hr incubation by submerging the fruits in water at 39.5°C for 2 min. This hot-water treatment at 39.5 to 42°C for 2 to 4 min has been the standard method for postharvest control of *Phytophthora* on citrus fruits for many years in California. Koltz and DeWolfe (1961) found that when lemons were submerged for 4 min in 48°C water, the temperature at a depth of 318 μ beneath the surface of the fruit reached 44°C. Both zoospores and hyphae of *Phytophthora* were killed by 1-min exposure to 44°C. In recent years a number of reports have indicated the feasibility of using heated water to control several postharvest diseases, namely *Rhizopus* and *Monilinia* on peaches and nectarines (Smith, 1962; Smith and Bassett, 1964), *Penicillium* and *Diplodia* on oranges (Smoot and Melvin, 1964), *Collectotrichum gloesporioides* on mangoes (Pennock and Maldonaldo, 1962; Smoot and Segall, 1963) and on papayas (Akamine and

Arisumi, 1953), *Gloeosporium* on apples (Burchill, 1964), and decay of pepper and cantaloupes (Johnson, 1968; Steward and Wells, 1970). Most of the recommendations for effective disease control have fallen in the range of 49 to 55.5°C for 2 to 5 min. It is reasonable to assume that these treatments were effective due to a direct effect of the heat on the pathogen (Klotz and DeWolfe, 1961; Smith and Bassett, 1964).

Relative Humidity and Air Velocity

All fruits and vegetables continue to lose vapor after they are harvested. If this water loss, or transpiration, is not retarded, the produce can rapidly become wilted, tough, or mushy, and consequently inedible. These symptoms of water loss become objectionable when vegetables have lost between 5% and 10% of their weight due to transpiration.

Water loss is rapid at low relative humidity and slow at high relative humidity. It is rapid at low relative humidity because the air in the room contains less water vapor than it can hold at the temperature of the room; thus, water vapor is readily transferred from the humid interior of the leaf or fruit to the relatively dry air. In contrast, if the relative humidity of the room is 100%, the air in the room and that in the vegetable are balanced in respect to moisture content, the gradient between the two is low, and moisture loss is nil. The amount of moisture the air can hold before it is saturated rises as the temperature rises; for example, more water is required to saturate air at 16°C than at 4.5°C. A room at 10°C and 90% R.H. is drier than a room at 4.5°C and 90% R.H., and dehydration would be more rapid in the room held at 10°C. Water has a greater tendency to evaporate as its temperature rises; thus, if a vegetable is at 10°C, the water it contains has a greater tendency to evaporate than if it were at 4.5°C. This means that water will be lost more rapidly from a warm than from a cool vegetable at a given air temperature and relative humidity.

The relative humidity recommended for each produce should be maintained in the storage room. High relative humidity (over 95%) is highly essential to avoid shrivelling, weight loss, and subsequent loss of flavor components. However, at such high relative humidity, prolific microbial growth takes place. Therefore, an appropriate amount of relative humidity is very important.

Minimizing the difference between the humidity inside the fruits and vegetables and that of the air and protecting the fresh produce from relatively dry air are two ways to reduce dehydration. This can be achieved by keeping the relative humidity of the air as close to saturation as possible by keeping the refrigerant close to the desired air temperature, which reduces condensing out the moisture in the air onto a cold surface. Moisture can be added to the air by water sprays or by fogging devices (Guillou, 1967). These are particularly useful during precooling when moisture loss is highest, and when dry containers literally can draw moisture from their contents.

As to the control of postharvest diseases of fruits and vegetables, a high relative humidity generally improves the fungicidal effectiveness of fumigants by increasing the degree of hydration of the pathogen. On the other hand, high humidity may reduce the practical efficiency of water-soluble fumigants such as sulfur dioxide and ammonia because they are strongly sorbed by surfaces in the fumigation chamber under high humidity conditions. Large increases in the applied dosage of the fumigants are necessary to counteract the effect of sorption at high humidities (Cant and Nelson, 1957; Nelson et al., 1964).

High humidity and relatively high temperatures favor the development of resistance of sweet potatoes and oranges to infection by pathogenic fungi. More commonly, however, high humidity favors the development of decay in fresh fruits (Loucks and Hopkins, 1946; Nelson et al., 1964).

As to circulation, the more rapidly the air moves across the surfaces of fruits and vegetables, the more rapidly water will be lost, unless the moving air is water-saturated. High velocity air causes rapid water loss because it continuously removes the extremely thin layer of saturated air that envelopes a leaf or fruit. Consequently, there exists a gradient which impels moisture to move from the humid interior of the tissue to the relatively dry room. Air movement should be just sufficient to effectively remove respiratory heat and ethylene from the product after it has been cooled to the temperature of the storage room.

Major Market Disorders and Their Control

Disorders are symptoms of disturbances in the normal metabolism or in the structural integrity of vegetables or fruits. Disorders, unlike diseases, are not caused by fungi, bacteria, or viruses. Disorders are the result of internal changes, such as senescence and external factors like unfavorable environment and physical injury.

Improper physical handling can result in injuries caused by impact, compression, abrasion, puncturing, tearing, or by two or more actions combined. Impact damage occurs when a product hits a surface with sufficient force to damage or even separate its cells. Compression, like impact, causes bruising or cracking, but occurs primarily during or after packing as a result of forcing too much product into too small a container. Abrasion can occur during harvest when roots or tubers are dug and conveyed at excessive speed, during packing when fruits are forced against rough containers, or during transit of slackly packed containers which permit the individual units to rub against each other or against the container surfaces. Punctures are sustained by cucumbers, eggplants, and some packs of tomatoes because their pedicels accidentally or intentionally remain attached. Tears can be sustained by leafy vegetables and, because of the tissue exposed, contribute to rapid dehydration, discoloration, or decay of the affected leaf.

Freezing injures vegetables and fruits by destruction of cells after the formation of ice crystals. The damaged, dead cells lose their resistance to dehydration and to microbial infections. Further, freeze-damaged tissue loses its normal rigidity and becomes mushy upon thawing, leading to the water-soaked appearance commonly associated with thawed vegetables. Freezing also may lead to development of strong off-odors upon cooking the damaged vegetable.

Greening of some normally white or cream-colored products is perhaps the only adverse effect of light on harvested vegetables. Greening can be particularly serious on potatoes because it develops at light intensities (about 20 fc) that are readily exceeded on the surface of produce counters in the food stores (Hardenburg, 1964). The greening of potatoes is due to the formation of chlorophyll (Larson, 1949). Greening can be reduced by filtering light through amber, yellow, green, or red sheets of cellophane, with amber the most effective. Greening can be prevented by displaying only one day's supply of sensitive crops, and keeping the reserve in opaque containers. Such rotation of stock is essential when food stores remain open around the clock (Hardenburg, 1964). Recent extensive research by Wu and Salunkhe (1972a, 1972b, 1972d) indicated that

greening of potato tubers can be completely prevented by treatment with hot wax or various oils.

Heat injury is the damage sustained when fruits and vegetables become substantially hotter than the surrounding air. Softening, discoloration, collapse, and eventual drying of affected tissue are the consequences of heat injury. These lesions are not only unsightly and inedible, but also offer avenues of entry for decay organisms. Sun scald can be minimized in some crops by maintaining good foliage cover over the fruits or when foliage becomes sparse, by a cover of nontoxic whitewash that is a water suspension of a proprietary material mainly consisting of aluminum silicate and of a spreader-sticker (Lipton and Matoba, 1971).

Ethylene and ammonia are the only volatiles likely to damage vegetables after harvest. If the ammonia concentration in the room exceeds 0.5%, the stored vegetables may be seriously injured in an hour or less (Ramsey and Butler, 1928). Ammonia injury manifests itself as a brown to black discoloration, particularly in the peripheral tissue that is naturally red, yellow, or brown. Normally white tissue, such as in white onion, turns greenish-yellow. After severe injury, deeper tissue may be affected and become mushy. Injury from ammonia can be avoided by proper maintenance of refrigeration equipment and by daily inspection of each storage room.

Yellowing of floret is a common disorder of broccoli and is a sign of senescence. Holding for prolonged periods or at temperatures that are too high leads to yellowing. Exposure to ethylene also accelerates yellowing, particularly at temperatures above 4.5°C. Head lettuce is afflicted by numerous disorders, some of which are difficult to distinguish from each other, or from diseases, because cells of lettuce often die in groups, turn some shade of red or brown, and collapse. Pink rib of lettuce is a pink discoloration of midribs and of smaller veins in severe instances (Marlott and Stewart, 1956). Pink rib has been increased substantially by low O_2 during one week at 10°C or one month at 2.3° or 5°C (Lipton, 1967, 1971). This can be minimized by precooling and shipment at 2.3°C or below. Rib discoloration of lettuce, also known as rib blight and brown rib, has yellow to black lesions in the midrib of cap leaves and in one or more of the next four or five younger leaves. Older and younger leaves occasionally may be affected. This can be prevented by misting lettuce during hot days because water cools the lettuce directly and indirectly via evaporative cooling (Jenkins, 1959).

Yellowing of fresh produce is a sign of senescence. It is accelerated by warm temperatures and the presence of ethylene in the air (Morris and Mann, 1946). Cucumbers held between 7° and 10°C yellow slowly. This fruit should not be stored with commodities that produce much ethylene, such as melons or apples. Vein tract browning is the unnetted, longitudinal, normally buff or green strips characteristic of the surface of some cantaloupes (Davis, 1970). The disorder can be delayed by precooling melons promptly to about 4.5°C.

Emergence of a sprout in an onion is the external indication that growth has occurred, and is preceded by the gradual elongation of the sprout within the bulb. Low temperature, 0°C or slightly below, effectively retards sprouting but does not prevent it in most onions if they are stored six months or longer. Sprouting also can be inhibited by preharvest application of various growth regulators, chiefly maleic hydrazide (Isenberg and Ang, 1963).

The disorder of potatoes, internal black spot, is a blackening of a small area of cortex tissue, although occasionally deeper tissues are affected. In early stages, the discoloration does not extend to the surface. However, in severe cases of black spot the symptoms may become strikingly evident. The external symptoms develop from desiccation of the affected internal tissue with subsequent depression of the tuber surface over the areas. Careful handling of the tubers at all stages of marketing, from harvesting through grading, transport, and destination handling, is the best protection against losses from black spot. Internal symptoms of potato blackheart are dark-gray to purplish or black discoloration. Freshly cut tissues may not show discoloration, but after brief exposures to air the typical symptoms appear. The affected tissues are usually in the central part of the tuber and sharply differentiated from healthy areas. In advanced stages, the affected tissues may dry out and separate to form cavities. External symptoms are not evident in the early stages, but advanced blackheart may show moist, discolored areas on the tuber surface. The most important step in avoidance of blackheart is adequate ventilation.

Internal breakdown of sweet potatoes some-times develops in storage. It is related to storage environment, cultivar, and intercellular space of the roots at harvest. Internal tissues of the affected roots are pithy or dry and spongy. In advanced cases, the roots are light in weight and compress easily. Accurate control of storage temperature, relative humidity, and air circulation will reduce severity of the disorder.

One of the most destructive storage troubles affecting the apple is the scald disorder. It is a physiological disorder that appears first on the unblushed or green side of the fruit. It has a characteristic burned or scald appearance. At first the discolored area is firm beneath the skin since the injury is confined to the skin. Later, the flesh immediately below the affected skin may become soft. Storage scald seems to result from accumulations of certain gases which the apple itself generates in its life processes (Brooks et al., 1919a, 1919b, 1923). This theory has since been widely accepted. Some cultivars of fruit, notably those with heavily pigmented skins, are resistant to the disorder. In susceptible cultivars, the scale areas tend to occur in the green rather than in the colored or russeted parts. Prestorage delay increases the incidence of scald. Today the biochemical causes still remain obscure and more of the methods of control, whether long established or of recent origin, have not been completely successful.

The only serious physiological disorder of bananas is chilling injury. In severe chilling, the green peel develops extensive subepidermal browning or blackening and the peel may become entirely black during ripening. Haard and Hultin (1969) found that the relative humidity must be at least 80% for normal ripening of "Cavendish" bananas. Below 80%, the banana failed to undergo a climacteric and did not ripen normally, eventually exhibiting symptoms characteristic of severe chilling.

The most commonly occurring physiological breakdown in oranges and grapefruits is pitting. The fruit may have areas of blotches of sunken spots. In oranges, the early and midseason cultivars are more affected than the late season ones. In grapefruit, the spots may first turn pink and then brown. Oleocellosis (oil spotting), which occurs in all citrus fruits, is especially severe in lemons and limes. Oranges may also suffer from brown stains or scald, which differ from pitting in appearance in that large areas are affected, and the affected area is seldom as dark as in the case of pitting. Water breakdown of citrus fruits involves symptoms very similar to those of freeze damage. The affected tissue has a water-soaked appearance. Ageing is a storage disorder where the rind around the stem becomes wilted and shrivelled, giving an unsightly appearance to the fruit. It is caused by loss of water from fruits, the collapse of the oil glands, and the subsequent death of the cells.

Chilling injury of pineapple is characterized by failure of the green shell to turn yellow; the yellow-shelled fruit turning a brown or dull color, a wilting, drying, and discoloration of crown leaves; and a breakdown of internal tissues, giving a watery appearance. Endogenous brown spot sometimes occurs simultaneously with chilling injury.

Scald in cherries is due to the localized translocation of pigment from the skin to the flesh as a consequence of bruising (Dekazos, 1960). Buch et al. (1961) found bruising of red tart cherries to be accompanied by demethylation of pectin.

Some of the aforestated physiological disorders in fruits and vegetables can be controlled by harvesting the produce at its proper maturity and promptly storing without damage in precisely controlled temperature, relative humidity, and air circulation.

CONCLUSIONS

Major influences on the ripening, senescence, and deterioration of fruits and vegetables are endogenous factors such as plant hormones produced within the plant body and exogenous factors which consist of environmental parameters such as microbial growth, temperature, relative humidity, air velocity, and atmospheric composition. A better and more effective utilization of fruit and vegetable crops can be achieved by careful manipulation of these factors. Application of appropriate chemicals may delay ripening and consequently extend the availability of fresh produce over a longer period of time. On the other hand, mature fruits and vegetables may be harvested at the "green" stage and stored under a controlled atmosphere with treatment by chemical agents to delay ripening and senescence and inhibit microbial decay, similarly extending their shelf life. Furthermore, CA-stored fruits and vegetables under precisely recommended temperature and humidity may subsequently be subjected to treatment with chemicals that can accelerate the ripening process and achieve uniform ripeness for immediate marketing.

Various kinds of nutritious fruits and vegetables are produced in great abundance. However, the losses of these highly perishable foods at the time of harvest and during storage are enormous. To cope with this problem practical regulation of ripening and senescence through chemical agents and economically feasible storage procedures is essential to minimize losses and assure sufficient supplies over a long period of time beyond harvest.

REFERENCES

Abdel-Gawad, H. S. and Romani, R. J., The regulatory effect of gibberellic acid and cytokinins on fruit maturation and ripening, Am. Soc. Hortic. Sci. 65th Ann. Marketing Prog., 1968 Abstr., 122.

Abdel-Kader, A. S., Morris, L. L., and Maxie, E. C., Effects of growth regulating substances on the ripening and shelf life of tomatoes, Hortic. Sci., 1, 90, 1966.

Abeles, F. B. and Rubenstein, B., Regulation of ethylene and leaf abscission by auxin, Plant Physiol., 39, 963, 1964.

Acosta, J. C. and York, T. L., Control of sprouting in onion bulbs with maleic hydrazide, Philipp. Agric., 40, 525, 1957.

Adams, R. E. and Tamuro, S. E., Treatment of field boxes for the control of postharvest rots of peaches and storage rots of apples, Plant Dis. Rep., 43, 396, 1959.

Akamine, E. K. and Arisumi, T., Control of postharvest storage decay of fruit of papaya (Carica papaya L.) with special reference to the effect of hot water, Proc. Am. Soc. Hortic. Sci., 61, 270, 1953.

Allen, F. W., The influence of growth regulator sprays on the growth, respiration and ripening of "Bartlett" pears, Proc. Am. Soc. Hortic. Sci., 62, 279, 1953.

Allen, F. W. and Claypool, L. L., Modified atmospheres in relation to the storage life of "Bartlett" pears, Proc. Am. Soc. Hortic. Sci., 52, 192, 1948.

Allen, F. W. and Smock, R. M., Carbon dioxide storage of apples, pears, plums, and peaches, *Proc. Am. Soc. Hortic. Sci.*, 35, 193, 1937.

Allentoff, M., Philips, W. R., and Johnston, F. B., A C^{14} study of CO_2 fixation in apples. I. The distribution of incorporated C^{14} in detached "McIntosh" apples, *J. Sci. Food Agric.*, 5, 231, 1954.

Ancheta, F. P., Ripening "Latundan" and "Lakatan" bananas with acetylene, Undergraduate thesis, University of Philippines, Lagung, Philippines, 1957.

Anderson, R. E., Parsons, C. S., and Smith, W. L., Jr., CA storage of Eastern-grown peaches and nectarines, USDA Marketing Res. Rep. No. 836, 1969.

Angell, F. F., The effect of Ethrel on ripening and quality of processing tomatoes, Abstr., *Hortic. Sci.*, 5, 318, 1970.

Audinay, A., Essal de controle artificiel de la maturation de l'ananas par l'ethrel, *Fruits* (Paris), 25(10), 695, 1970.

Ayres, J. C. and Denisen, E. L., Maintaining freshness of berries using selected packaging materials and antifungal agents, *Food Technol.*, 12, 562, 1958.

Ayres, J. C., Kraft, A. A., and Pierce, L. C., Delaying spoilage of tomatoes, *Food Technol.*, 18, 1210, 1964.

Barger, W. R., Wiant, J. S., Pentzer, W. T., Ryall, A. L., and Dewey, D. H., A comparison of fungicidal treatments for the control of decay in California cantaloups, *Phytopathology*, 38, 1019, 1948.

Bate-Smith, E. C., Flavonoid compounds in foods, *Adv. Food Res.*, 5, 261, 1954.

Batjer, L. P. and Moon, M. H., Effect of NAA spray on maturity of apples, *Proc. Am. Soc. Hortic. Sci.*, 46, 113, 1945.

Batjer, L. P., Siegelman, H. W., Rogers, B. L., and Gerhart, F., Results of 4 years' test of the effect of 2,4,5-TP on maturity and fruit drop of apple in the Northwest, *Proc. Am. Soc. Hortic. Sci.*, 64, 215, 1954.

Beccari, F., Banana conditioning during transport without refrigeration, *Proc. Conf. Trop. Subtrop. Fruits* (London), 1969, 93.

Becker, R. F., Goodman, R. N., and Goldberg, H. S., Post-harvest treatment to control strawberry fruit rot, *Plant Dis. Rep.*, 42, 1066, 1958.

Beraha, L., Ramsey, G. B., Smith, M. A., and Wright, W. R., Gamma radiation for possible control of post-harvest diseases of apples, strawberries, grapes, and peaches, *Phytopathology*, 47, 4, 1957.

Beraha, L., Smith, M. A., and Wright, W. R., Gamma radiation dose response of some decay pathogens, *Phytopathology*, 50, 474, 1960.

Berard, J. E., Memoire sur la maturation des fruits, *Ann. Chim. Phys.*, 16, 178, 1821.

Besser, T. and Kanner, A., Extending storage life of mushrooms with carbon monoxide, *Proc. 18th Intl. Hortic. Cong.*, 1, 30, 1970.

Bezlur, R. M. and Joslyn, M. A., Die phenolase and peroxydase de pflaumen, *Hortic. Abstr.*, 24, 2303, 1954.

Blackman, R. R. and Parija, P., Analytic studies in plant respiration. II. Formation of catalytic system for the respiration of apples and relation to oxygen, *Proc. R. Soc. Lond. B. Biol. Sci.*, 103, 412, 1928.

Boe, A. A., Some biochemical changes occurring in tomato fruit ripened by several treatments, Ph.D. dissertation, Utah State University, Logan, Utah, 1966.

Bondad, N. D., Effects of Ethrel on the ripening of banana fruits, Undergraduate thesis, University of the Philippines, La Laguna, Philippines, 1971.

Bondad, N. D., Mendoza, R. C., and Pantastico, E. B., Postharvest ripening of tomato fruits with 2-chloroethylphosphonic acid and calcium carbide, *Proc. Crop. Sci. Soc. Philipp.*, 2, 79, 1971.

Bondad, N. D. and Pantastico, E. B., Note: Response of tomato fruits to acetylene and calcium carbide treatments, *Philipp. Agric.*, 55, 333, 1971.

Borgstrom, G., *Principles of Food Science*, Vol. 1 and 2, Borgstrom, G., Ed., The MacMillan Co., New York, 1968.

Bottini, E., Sulla conservazione della frutta. Ricerche sperimentali esequite sulle Arancie, *Ann. Chim. Appl.*, 17, 129, 1927.

Boyle, F. P., Kertesz, Z. I., Clegg, R. E., and Conner, M. A., Jr., Effects of ionizing radiation on plant tissues. II. Softening of different varieties of apples and carrots by gamma rays, *Food Res.*, 22, 89, 1957.

Brendler, R. A., Field ripening peppers with Ethrel, 1964, Agric. Ext. Univ. Calif. (Ventura), February, 1970.

Brooks, C., Modified atmospheres for fruits and vegetables in storage and transit, *Refrig. Eng.*, 40, 233, 1940.

Brooks, C. and Cooley, J. S., Effect of temperature, aeration, and humidity on Jonathan-spot and scald of apples in storage, *J. Agric. Res.*, 11, 306, 1917.

Brooks, C., Cooley, J. S., and Fisher, D. F., Relation of character of fruit to scald development, *J. Agric. Res.*, 16, 195, 1919a.

Brooks, C., Cooley, J. S., and Fisher, D. F., Nature and control of apple-scald, *J. Agric. Res.*, 18, 211, 1919b.

Brooks, C., Cooley, J. S., and Fisher, D. F., Oiled wrappers, oils and waxes in the control of apple scald, *J. Agric. Res.*, 26, 513, 1923.

Brownell, L. E., Gustafson, F. G., Nehemias, J. V., Isleib, D. R., and Hooker, W. J., Storage properties of gamma irradiated potatoes, *Food Technol.*, 11, 306, 1957.

Buch, M. L., Satori, K. G., and Hills, C. H., The effects of bruising and aging on the texture and pectic constituents of canned red tart cherries, *Food Technol.*, 15, 526, 1961.

Bullock, R. M. and Rogers, B. L., Color development in apples as influenced by pre-harvest sprays of 2,4,5-TP, *Wash. State Hortic. Assoc.*, 47, 21, 1952.

Burchill, R. T., Hot water as a possible post-harvest control of *Gloeosporium* rots of stored apples, *Plant Pathol.*, 13, 106, 1964.

Burg, S. P. and Burg, E. A., Fruit storage at sub-atmospheric pressures, *Science,* 153, 314, 1966.

Burg, S. P. and Burg, E. A., Molecular requirements for the biological activity of ethylene, *Plant Physiol.,* 42, 144, 1967.

Burgheimer, F., McGill, J. N., Nelson, A. I., and Steinberg, M. P., Chemical changes in spinach stored in air and controlled atmosphere, *Food Technol.,* 21, 1273, 1967.

Burton, W. G. and Hannan, R. S., Use of gamma radiation for preventing the sprouting of potatoes, *J. Sci. Food Agric.,* 8, 707, 1957.

Cant, R. R. and Nelson, K. E., Factors affecting the concentration of the sulfur dioxide in fumigation atmospheres for table grapes, *Proc. Am. Soc. Hortic. Sci.,* 69, 240, 1957.

Cappellini, P. and Fideghelli, C., Effect of 2,4,5-trichlorophenoxypropionic acid on enzymatic activity of leaves and fruits in peach and table grape, *Proc. 18th Intl. Hortic. Cong.,* 1, 174, 1970.

Cappellini, R. A. and Stretch, A. W., Control of postharvest decays of peaches, *Plant Dis. Rep.,* 46, 31, 1962.

Cardinell, H. A. and Barr, C. G., Postharvest tests with peaches to reduce spoilage, *Mich. State Univ. Agr. Exp. Stn. Q. Bull.,* 35, 39, 1952.

Celestino, A. F., The effects of irrigation, nitrogen fertilization and maleic hydrazide spray on the yield, composition and storage behavior of bulbs of two onion varieties, *Phil. Agric.,* 44, 479, 1960.

Celestino, A. F. and Ventura, A. F., The influence of maleic hydrazide on the storing quality of the Irish potato tubers, Proj. Rep. of the Veg. Crops Section, UPCA, Coll., Laguna, Philippines, 1960.

Chauhan, K. S. and Parmar, C., Degreening of "Mosambi" orange with Ethrel (2-chloroethylphosphonic acid), *Proc. 3rd Intl. Symp. Trop. Subtrop. Hortic.,* 1972, 82.

Chawan, T. and Pflug, I. J., Controlled atmosphere storage of onions, *Mich. State Univ. Agr. Exp. Stn. Q. Bull.,* 50, 449, 1967.

Chhongkar, V. A. and Sengupta, B. M., Effects of plant growth regulators on growth, yield and quality of some important vegetables, *Proc. 3rd Intl. Symp. Trop. Subtrop. Hortic.,* 1972, 47.

Chiang, M. N., Studies on the removal of ethylene from the controlled atmosphere storage of bananas, *Coll. of Agric. Natl. Taiwan Univ. Special Publ.,* No. 21, 1968.

Chiarappa, L., Eckert, J. W., and Kolbezen, M. J., Effect of dibromotetrachlorolthane and other chemicals on Botrytis decay of Emperor grapes in storage, *Am. J. Enol. Vitic.,* 13, 83, 1962.

Childs, J. F. L. and Siegler, E. A., Controlling orange decay, *Ind. Eng. Chem.,* 2nd ed., 38, 82, 1946.

Chourdri, R. S. and Bhatnagar, V. B., Effect of maleic hydrazide on the keeping quality of turnips (*Brassica rapa*), *Ind. J. Hortic.,* 12, 1, 1955.

Claypool, L. L. and Allen, F. W., Modified atmosphere in relation to the transportation of deciduous fruits, *Proc. Am. Soc. Hortic. Sci.,* 49, 92, 1947.

Claypool, L. L. and Allen, F. W., Carbon dioxide production of deciduous fruits held at different oxygen levels during transit period, *Proc. Am. Soc. Hortic. Sci.,* 51, 103, 1948.

Claypool, L. L., Pangborn, R. M., and Shaver, J. F., Influence of controlled atmosphere storage on quality of canned apricots, *Proc. 17th Intl. Hortic. Cong. Abstr.,* 1, 659, 1966.

Cloutier, J. A. R., Xoc, C. E., Manson, J. M., Clay, M. C., and Johnson, L. E., Effect of storage on carbohydrate content of two varieties of potatoes grown in Canada and treated with gamma radiation, *Food Res.,* 24, 659, 1959.

Coggins, C. W. and Eaks, I. L., Gibberellin research on "Navel" oranges, *Calif. Citrogr.,* 52, 475, 1967.

Coggins, C. W. and Hield, H. Z., "Navel" orange fruit response to potassium gibberellate, *Proc. Am. Soc. Hortic. Sci.,* 81, 227, 1962.

Coggins, C. W., Hield, H. Z., and Boswell, S. B., The influence of potassium gibberellate on "Lisbon" lemon trees and fruit, *Proc. Am. Soc. Hortic. Sci.,* 76, 199, 1960.

Coggins, C. W., Hield, H. Z., and Burns, R. M., Influence of potassium gibberellate on grapefruit trees and fruit, *Proc. Am. Soc. Hortic. Sci.,* 81, 223, 1962.

Cooper, G. M. and Salunkhe, D. K., Effects of gamma-radiation, chemical, and packaging treatments on refrigerated life of strawberries and sweet cherries, *Food Technol.,* 17(6), 123, 1963.

Cooper, W. C. and Henry, W. H., Abscission chemicals in relation to citrus fruit harvest, *J. Agric. Food Chem.,* 19, 559, 1971.

Cooper, W. C., Rasmussen, G. K., and Smoot, J. J., Induction of degreening of tangerines by preharvest applications of ascorbic acid and other ethylene-releasing agents, *Citrus Ind.,* 49(10), 25, 1968.

Copisarow, M., Maleic acid for plant spraying, *J. Chem. Ind.,* 55, 746, 1936a.

Copisarow, M., The metabolism of fruits and vegetables in relation to their preservation, *J. Pomol. Hortic. Sci.,* 14, 9, 1936b.

Couey, H. M., Effect of temperature and modified atmosphere on the storage life, ripening behavior, and dessert quality of "Eldorado" plums, *Proc. Am. Soc. Hortic. Sci.,* 75, 207, 1960.

Couey, H. M., Modified atmosphere storage of "Nubiana" plums, *Proc. Am. Soc. Hortic. Sci.,* 86, 166, 1965.

Cox, R. S., A preliminary report on disease of lettuce in the Everglades and their control, *Plant Dis. Rep.,* 39, 421, 1955.

Crandall, P. C., Relation of preharvest sprays of maleic hydrazide to the storage life of "Delicious" apples, *Proc. Am. Soc. Hortic. Sci.,* 65, 71, 1955.

Crane, J. C., Marie, N., and Nelson, M. M., Growth and maturation of fig fruits stimulated by 2-chloroethylphosphonic acid, *J. Am. Soc. Hortic. Sci.,* 95, 367, 1970.

Crocker, W., Hitchcock, A. E., and Zimmerman, P. W., Similarities in the effects of ethylene and the plant auxins, *Contrib. Boyce Thompson Inst.,* 7, 231, 1935.

Daito, H. and Hirose, K., Studies on acceleration of coloring or degreening of citrus fruits. II. Effects of Ethrel (ethylene-releasing compound) on the acceleration of coloring and carotenoid pattern of the "Natsudaidai" (*Citrus natsudaidai* Hayata), *Bull. Hortic. Res. Stn. Ser. B. (Okitsu),* No. 10, 35, 1970.

Damigella, P., Experiments on the effectiveness of 2,4-D and 2,4,5-T on mandarins, *Tech. Agric.,* 14, 430, 1962.

Davis, R. M., Jr., Vein tracts, not sutures, in cantaloupe, *Hortic. Sci.,* 5:86, 1970.

Dedolph, R. R., Wittwer, S. H., and Tuli, V., Senescence inhibition and respiration, *Science,* 134, 1075, 1961.

Dedolph, R. R., Wittwer, S. H., Tuli, V., and Gilbart, D., Effect of N^6-benzylaminopurine on respiration and storage behavior of broccoli (*Brassica oleracea* var. "Italica"), *Plant Physiol.,* 37, 509, 1962.

Dekazos, E. D., Anthocyanin in red tart cheries as related to anaerobiosis and scald, *J. Food Sci.,* 31, 226, 1966.

Dennis, D. T., Upper, C. C., and West, C. A., An enzymatic site of inhibition of gibberellin biosynthesis by Amo-1618 and other plant growth retardants, *Plant Physiol.,* 40, 948, 1965.

Dennis, F. G., Jr., Wilczynski, H., Dela Guardia, M., and Robinson, R. W., Ethylene levels in tomato fruits following treatment with ethrel, *Hortic. Sci.,* 5, 168, 1970.

Desrosier, N. W., *The Technology of Food Preservation,* Desrosier, N. W., Ed., The Avi Publishing Co., Inc., Westport, Conn., 1970.

Dewey, D. H. and Wittwer, S. H., Chemical control of top growth of prepackaged radishes, *Proc. Am. Soc. Hortic. Sci.,* 66, 322, 1955.

DiMarco, G. R. and Davis, B. H., Prevention of decay of peaches with post-harvest treatments, *Plant Dis. Rep.,* 41, 284, 1957a.

DiMarco, G. R. and Davis, B. H., Prevention of decay on strawberries with post-harvest treatments, *Plant Dis. Rep.,* 41:460, 1957b.

Do, J. Y., Salunkhe, D. K., Sisson, D. V., and Boe, A. A., Effects of hydrocooling, chemical, and packaging treatments on refrigerated life and quality of sweet cherries, *Food Technol.,* 20, 115, 1966.

Domenico, J. A., Rahman, A. R., and Westcott, D. E., Effects of fungicides in combination with hot water and wax on the shelf life of tomato fruit, *J. Food Sci.,* 37, 957, 1972.

Dostal, H. C. and Leopold, A. C., Gibberellin delays ripening of tomatoes, *Science,* 158, 1579, 1967.

Dostal, H. C. and Wilcox, G. E., Chemical acceleration of ripening of field grown tomatoes, Abstr., *Proc. 18th Int'l. Hortic. Cong.,* 1, 188, 1970.

Duncan, D. T., Hooker, W. J., and Heiligman, F., Storage rot susceptibility of potato tubers exposed to minimum sprout inhibiting levels of ionizing radiation, *Food Technol.,* 13, 159, 1959.

Dutcher, R., Flower induction in "Carabao" mango, *Univ. Philipp.-Cornell Rep. Coll.,* Laguna, 1971.

Eaks, I. L., Eckert, J. W., and Roistacher, C. M., Ammonia gas fumigation for control of *Rhizopus* rot on peaches, *Plant Dis. Rep.,* 42, 846, 1958.

Eaves, C. A., Dry scrubber for the removal of carbon dioxide from controlled atmosphere storage for apples, *Annu. Rep. Nova Scotia Fruit Growers Assoc.,* 95, 105, 1958.

Eaves, C. A., Semi-automatic dry scrubber for CA apple storage, *Can. Dept. Agric. Res. Farmers,* 7(11), 12, 1962.

Eckert, J. W., The place of the fungistat, *Calif. Citrogr.,* 44, 281, 1959.

Eckert, J. W. and Kolbezen, M. J., 2-Aminobutane salts for control of post-harvest decay of citrus, apple, pear, peach, and banana fruits, *Phytopathology,* 54, 978, 1964.

Eckert, J. W., Kolbezen, M. J., and Garber, M. J., Evaluation of ammonia generating formulations for control of citrus fruit decay, *Phytopathology,* 53, 140, 1963.

Edgerton, L. J. and Hoffman, M. B., Inhibition of fruit drop and color stimulation with N-dimethylamino succinamic acid, *Nature,* 209, 314, 1966.

Eilati, S. K., Goldschmidt, E. E., and Monselise, S. P., Hormonal control of color changes in orange peel, *Experientia,* 25, 209, 1969.

El-Mansy, H. I., Effects of pre- and post-harvest applications of 6-furfuryl-aminopurine and N^6-benzyladenine on physio-chemical changes in lettuce (*Lactuca sativa* L.), Master's thesis, Utah State University, Logan, Utah, 1964.

El-Mansy, H. I., Salunkhe, D. K., Hurst, R. L., and Walker, D. R., Effects of pre- and post-harvest applications of 6-furfurylaminopurine and N^6-benzyladenine on physiological and chemical changes in lettuce, *Hortic. Res.,* 7, 81, 1967.

El-Tabey, A. M. and Cruess, W. V., The oxidase of the apricot, *Plant Physiol.,* 24, 307, 1949.

Emmert, F. H. and Southwick, F. W., The effect of maturity, apple emanations, waxing and growth regulators on the respiration and color development of tomato fruits, *Proc. Am. Soc. Hortic. Sci.,* 63, 393, 1953.

Farkas, A., The practical application of impregnated wrappers against fungal decay of citrus fruit, *Hadar,* 11, 261, 1938.

Farkas, A., Control of wastage of citrus fruit by impregnated wrappers, on a commercial scale, *Hadar,* 12, 227, 1939.

Fawcett, H. S., Packinghouse control of brown rot, *Calif. Citrogr.,* 7, 233, 1922.

Fisher, D. V., Mealiness of quality in "Delicious" apples as affected by certain orchard conditions and storage techniques, *Proc. Am. Soc. Hortic. Sci.,* 40, 128, 1942.

Forsyth, F. R., Eaves, C. A., and Lockhart, C. L., Controlling ethylene levels in the atmosphere of small containers of apples, *Can. J. Plant Sci.,* 47, 717, 1967.

Frenkel, C., Involvement of peroxidase and indole-3-acetic acid oxidase isozymes from pear, tomato, and blueberry fruit in ripening, *Plant Physiol.*, 49, 757, 1972.

Frenkel, C. and Dyck, R., Auxin inhibition of ripening in "Bartlett" pears, *Plant Physiol.*, 51, 6, 1973.

Frenkel, C., Klein, I., and Dilley, D. R., Methods for the study of ripening and protein synthesis in intact pome fruits, *Phytochemistry*, 8, 945, 1969.

Frenkel, C. and Patterson, M. E., The effect of carbon dioxide on succinic dehydrogenase in peas during cold storage, *Hortic. Sci.*, 4:165, 1969.

Fuchs, Y. and Cohen, A., Degreening of citrus fruit with ethrel (Amchem. 66-329), *J. Am. Soc. Hortic. Sci.*, 94(6), 617, 1969.

Fulton, S. H., The cold storage of small fruits, USDA Bureau of Plant Industry Bull. No. 108, 1907.

Garrison, S. A., Stimulation of tomato ripening by Amchem, 66, 329, Abstr., *Hortic. Sci.*, 3, 122, 1968.

Gerhardt, F. and Allmendinger, D. F., The influence of α-naphthalene acetic acid spray on the maturity and storage physiology of apples, pears, and sweet cherries, *Proc. Am. Soc. Hortic. Sci.*, 46, 113, 1945.

Gerhardt, R. and Ezell, B.D., Effect of CO_2 storage on "Bartlett" pears under simulated transit conditions, *J. Agric. Res.*, 56, 121, 1938.

Goddard, D. R. and Meeuse, B. J. D., Respiration of higher plants, *Annu. Rev. Plant Physiol.*, 1, 207, 1950.

Godfrey, G. H. and Ryall, A. L., The control of transit and storage decays in Texas lemons, *Tex. Agric. Exp. Stn. Bull.*, 701, 1, 1948.

Gooding, H. J. and Campbell, J. S., Improvement of sweet potato storage by cultural and chemical means, *Emp. J. Exp. Agric.*, 32, 65, 1964.

Gortner, W. A., U.S. Patent 334,697, 1963.

Gortner, W. A., Relation of chemical structure to plant growth regulator activity in the pineapple plant: retarding senescence of pineapple fruit with application of 2,4,5-trichlorophenoxy acetic acid and α-naphthalene acetic acid, *J. Food Sci.*, 34, 577, 1969.

Gortner, W. A. and Leeper, R., Studies on the relation of chemical structure to plant growth regulator activity in the pineapple fruits. V. Postharvest delay of senescence of pineapple fruit, *Bot. Gaz.*, 130, 87, 1969.

Groeschel, E. C., Nelson, A. I., and Steinberg, M. P., Changes in color and other characteristics of green beans stored in controlled refrigerated atmosphere, *J. Food Sci.*, 31, 436, 1966.

Guillou, R., Problems in perishables handling — the engineer's view, Proc. Fruit and Vegetable Perishable Handling Conference, University of Calif., Davis, Calif., 5.

Gull, D. D., Ripening behavior and edible quality of tomato treated with Amchem 68-62 (Ethrel), *Proc. Fla. Sta. Hortic. Sci.*, 81, 214, 1969.

Haard, N. F. and Hultin, H. O., Abnormalities in ripening and mitochondrial succinoxidase resulting from storage of pre-climacteric banana fruit at low relative humidity, *Phytochemistry*, 8, 2149, 1969.

Halevy, A. H., Dilley, D. R., and Wittwer, S. H., Senescence inhibition and respiration induced by growth retardants and N^6-benzyladenine, *Plant Physiol.*, 41, 1085, 1966.

Halevy, A. H. and Wittwer, S. H., Chemical regulation of leaf senescence, *Mich. Agric. Exp. Stn. Q. Bull.*, 48, 30, 1965.

Halevy, A. H. and Wittwer, S. H., Increasing the longevity of perishable vegetables, mushrooms, and cut flowers by treatment with growth retardants, *Proc. Am. Soc. Hortic. Sci.*, 88, 582, 1966.

Hall, E. G. and Long, J. K., Citrus wastage investigations, *Agric. Gaz. N. S. W.*, 61, 631, 1950.

Hall, W. C. and Morgan, P. W., Auxin-ethylene interrelationships, in *Regulateurs Naturels de la Croissance Vegetable*, Nitsch, J. P., Ed., Centre National de la Recherche Scientifique, 1964, 727.

Haller, M. H., Handling, transportation, storage, and marketing of peaches, *U.S.D.A. Biograph. Bull.*, 21, 1, 1952.

Haller, M. H., Effect of 2,4,5-trichlorophenoxypropionic acid on maturity and storage quality of apples, *Proc. Am. Soc. Hortic. Sci.*, 64, 222, 1954.

Hannan, R. A., The preservation of food with ionizing radiations, *Food Sci. Abstr.*, 26, 121, 1954.

Hardenburg, R. E., Greening of potatoes during marketing. A review, *Am. Potato J.*, 41, 215, 1964.

Hardenburg, R. E., Schomer, H. A., and Uota, M., Polyethylene film for fruit, *Mod. Packag.*, 31, 135, 1958.

Harel, E., Mayer, A. M., and Shain, Y., Catechol oxidation, endogenous substrates and browning in developing apples, *J. Sci. Food Agric.*, 17, 389, 1966.

Harvey, J. M., Effects of frequency of sulfur dioxide fumigation during storage on decay and fumigation injury in Emperor grapes, *Phytopathology*, 46, 690, 1956.

Harvey, J. M. and Pentzer, W. T., Market diseases of grapes and other small fruits, U.S.D.A. Agric. Hdbk., 189, 1, 1960.

Hatton, T. T., Responses of the "Delicious" apples to sprays of 2,4,5-trichlorophenoxypropionic acid, *Proc. Am. Soc. Hortic. Sci.*, 65, 59, 1955.

Hatton, T. T., Effects of waxes and 2,4,5-trichlorophenoxyacetic acid as postharvest treatments of "Persian" limes, *Proc. Fla. State Hortic. Soc.*, 71, 312, 1958.

Heiberg, B. C. and Ramsey, G. B., Fungistatic action of biphenyl on some fruit and vegetable pathogens, *Phytopathology*, 36, 887, 1946.

Hield, H. S. and McCarthy, C. C., 2,4,-D control of preharvest drop of "Navel" oranges in Tulare county, *Univ. Calif. Dept. Hortic. Newsletter*, 10, 1, 1956.

Higdon, R. J., The effects of 2,4,5-trichlorophenoxyacetic acid on the development and ripening of peach fruits, *Proc. Am. Soc. Hortic. Sci.*, 58, 73, 1951.

Hill, G. R., Respiration of fruits and growing plant tissue in certain cases, with reference to ventilation and fruit storage, *Cornell Univ. Agric. Exp. Stn. Bull.*, No. 30, 1913.

Hirai, J., Hirata, N., and Tada, H., Effect of oleification on hastening the maturity of fig fruits. I. On the time of application and kind of oils. *J. Jap. Soc. Hortic. Sci.*, 35, 354, 1966.

Hirai, J. and Horiuchi, S., Effect of oleification on hastening the maturity of fig fruits. II. Respiration and changes in the concentrations of metabolic substances in the treated fig fruit with ripe seed oil, *J. Jap. Soc. Hortic. Sci.*, 36, 36, 1967a.

Hirai, J. and Horiuchi, S., Effect of oleification on hastening the maturity of fig fruits. III. Promoting effect of fatty acids and glycerine on fruit maturation, *J. Jap. Soc. Hortic. Sci.*, 36, 147, 1967b.

Hirai, J. and Horiuchi, S., Effect of oleification on hastening the maturity of fig fruits. IV. Respiration and changes in the concentrations of metabolic substances in fruits treated with fatty acid or glycerine, *J. Jap. Soc. Hortic. Sci.*, 36, 268, 1967c.

Hirai, J. and Horiuchi, S., Effect of oleification on hastening the maturity of fig fruits. V. Effect of metabolic products of fatty acid oxidation on fruit maturity, *J. Jap. Soc. Hortic. Sci.*, 36, 380, 1967d.

Hirai, J. and Horiuchi, S., Effect of oleification on hastening the maturity of fig fruits. VI. Respiration and changes in the concentrations of metabolic substances in fruits treated with products of fatty acids—oxidation, acetaldehyde or ethylene, *J. Jap. Soc. Hortic. Sci.*, 37, 20, 1968.

Hopkins, E. F. and Loucks, K. W., Studies on thiourea for stem-end rot control, *Fla. Univ., Agric. Exp. Stn. Annu. Rep.*, 1946, 164.

Hopkins, E. F. and Loucks, K. W., A curing procedure for the reduction of mold decay in citrus fruits, *Fla. Univ., Agric. Exp. Stn. Bull.*, 450, 1-26, 1948.

Hsieh, P. T., Yang, C. T., Tsai, S. S., Yang, C. C., and Hsiao, F. H., Study on storage of banana fruit. The effect of the storage condition sealed with plastic bag and sprayed with gibberellin solution on the main constituents of banana fruit, *J. Chin. Agric. Chem. Soc.*, 1, 1, 1967.

Huelin, F. E. and Tindale, G. B., The gas storage of "Victorian" apples, *J. Dept. Agric. Victoria*, 45(2), 74, 1947.

Hulme, A. C., Carbon dioxide injury and the presence of succinic acid in apples, *Nature*, 178, 218, 1956.

Hulme, A. C., Some aspects of the biochemistry of apples and pear fruit, *Adv. Food Res.*, 8, 297, 1958.

Hummel, C. E. and Stoddard, E. S., Methods of improving food preservation in home refrigerators, *Refrig. Eng.*, 65(6), 33, 1957.

Isaac, W. E., The use of maleic acid as a ripening inhibitor for deciduous fruits, *Low Temp. Res. Lab. Capetown Ann. Rep. for 1935-36*, 1937, 199.

Isaac, W. E., Further experiments concerning the use of maleic acid as a fruit ripening inhibitor, *Low Temp. Res. Lab. Capetown Ann. Rep. for 1936-37*, 1938, 177.

Isenberg, F. M., The use of maleic hydrazide on onions, *Proc. Am. Soc. Hortic. Sci.*, 68, 343, 1956.

Isenberg, F. M. and Ang, J. K., Northern-grown onions. Curing, storing, and inhibiting sprouting, *Cornell Ext. Bull.*, 1116, 1963.

Isenberg, F. M. and Sayles, R. M., Modified atmosphere storage of Danish cabbage, *J. Am. Soc. Hortic. Sci.*, 94, 447, 1969.

Iwahori, S., Ben-Yehoshua, S., and Lyons, J. M., Effect of 2-chloroethanephosphonic acid on tomato fruit development and maturation, *Bioscience*, 19, 49, 1968.

Iwahori, S. and Lyons, J. M., Accelerating tomato fruit maturity with Ethrel, *Calif. Agric.*, 23(6), 17, 1969.

Jarvis, W. R., The preservation of fruit chip baskets with copper-8-quinolinate, *Plant Pathol.*, 9, 150, 1960.

Jenkins, J. M., Jr., Brown rib of lettuce, *Proc. Am. Soc. Hortic. Sci.*, 74, 587, 1959.

Johnson, B., Mayer, M. M., and Johnson, D. K., Isolation and characterization of peach tannins, *Food Res.*, 16, 169, 1951.

Johnson, F., Effects of electromagnetic waves on fungi, *Phytopathology*, 22, 277, 1932.

Johnson, H. B., Heat and other treatments for cantaloups and peppers, *United Fresh Fruit and Veget. Assoc. Yearbook*, 1968, 51.

Iwahori, S. and Lyons, J. M., Maturation and quality of tomatoes with preharvest treatments of 2-chloroethylphosphonic acid, *J. Am. Soc. Hortic. Sci.*, 95, 88, 1970.

Jadhav, S. J., Patil, B. C., and Salunkhe, D. K., Control of potato greening under hypobaric storage, *Food Eng.*, 45(8), 111, 1973.

Jarrett, L. D. and Gathercole, F. J., Trials with aminobutane as alternative to SOPP reaffirms desirability of continued treatments, *Citrus News*, 40, 126, 1964.

Joslyn, M. A., Enzyme catalyzed oxidative browning of fruit products, *Adv. Food Res.*, 3, 1, 1951.

Kabachnik, M. I. and Rossiiskaya, P. A., Organophosphorous compounds. I. Reaction of ethylene oxide with phosphorous trichloride. Otdel Khim. Nauk 295, *C.A.*, 42, 7241f, 1946.

Kader, A. A., Brecht, P. E., Woodruff, R., and Morris, L. L., Influence of carbon monoxide, carbon dioxide, and oxygen levels on brown stain, respiration rate, and visual quality of lettuce, *J. Am. Soc. Hortic. Sci.*, 98, 485, 1973.

Karchi, Z., Effects of 2-chloroethanephosphonic acid on flower types and flowering sequence in muskmelons, *J. Am. Soc. Hortic. Sci.*, 95, 515, 1970.

Karnik, V. V., Selected physio-chemical, microbiological and agronomical studies on the controlled atmosphere storage of sugar beet (*Beta vulgaris*) roots, Ph.D. dissertation, Utah State University, Logan, Utah, 1970.

Karnik, V. V., Salunkhe, D. K., Olson, L. E., and Post, F. J., Physiochemical and microbiological effect of controlled atmosphere on sugar beet, *J. Am. Soc. Sugar Beet Technol.*, 16, 156, 1970.

Kasmire, R. F., Rappaport, L., and May, D., Effects of 2-chloroethylphosphonic acid on ripening of cantaloups, *J. Am. Soc. Hortic. Sci.*, 95, 134, 1970.

Kato, T., Mechanism of retardation of sprouting and occurrence of functional disorder in onion bulbs applied with maleic hydrazide, *Engei Gakkai Zasshi*, 38(2), 178; *C.A.*, 74, 98656a, 1970.

Katsumi, M., Physiological effects of kinetin. Effect of kinetin on the elongation, water uptake and oxygen uptake of etiolated pea stem sections, *Physiol. Plant*, 16, 66, 1963.

Kaufman, J. and Ringel, S. M., Tests of growth regulators to retard yellowing and abscission of cauliflower, *Proc. Am. Soc. Hortic. Sci.*, 78, 349, 1961.

Kenworthy, A. L., Fruit maturation and quality of sour and sweet cherry and other stone fruits as influenced by Alar (Succinic acid 2,2-dimethyl hydrazide), *Proc. 18th Int'l. Hortic. Cong.*, 1, 4, 1970.

Kidd, F. and West, C., The gas storage of fruit. II. Optimum temperatures and atmospheres, *J. Pomol. Hortic. Sci.*, 8, 67, 1930.

Kidd, F., West, C., and Kidd, M. N., Gas storage of fruit, *G. B. Dept. Sci. Ind. Res. Food Invest. Spec. Rep.*, 20, 1, 1927.

Kitagawa, H., Suguiura, A., and Sugiyama, M., Effects of gibberellin spray on storage quality of "Kaki," *Hortic. Sci.*, 7(2), 59, 1966.

Klater, E. and Rudich, J., Enhanced ripening of processing tomatoes by application of Ethrel, Abstr., *Proc. 18th Int'l. Hort. Cong.*, 1, 187, 1970.

Klotz, L. J. and DeWolfe, T. A., Limitations of hot water immersion treatment for the control of *Phytophthora* brown rot of lemons, *Plant Dis. Rep.*, 45, 264, 1971.

Krishnamurthy, S. and Subramanyam, H., Effect of maleic hydrazide and 2,4,5-trichlorophenoxypropionic acid on ripening and quality of mango fruit, *Pesticide Sci.*, 1, 63, 1970.

Kung, H. C., Gaden, E. L., and King, C. G., Vitamins and enzymes in milk: Effects of gamma radiation on activity, *J. Agric. Food Chem.*, 1, 142, 1953.

Lakshminarayana, S. and Subramanyam, H., Effect of preharvest spray of maleic hydrazide and isopropyl n-phenyl-carbamate on sapota (*Archras sapota* Linn.). II, *J. Food Sci. Technol.*, 4, 70, 1967.

Lakshminarayana, S., Subramanyam, H., and Surendranath, V., Effect of preharvest spray of growth regulators on the size, composition, and storage behavior of sapota (*Archras sapota* Linn.). I, *J. Food Sci. Technol.*, 4, 66, 1967.

Lampe, C. H., Response of tomato fruits to certain growth regulators with emphasis on pectolytic enzymes, cellulose and ethylene, *Diss. Abst.*, 32, 1308B, 1971.

Lannert, J. W., Tectrol atmospheric control, *Agric. Eng.*, 45, 318, 1964.

Larson, E. C., Investigations on cause and prevention of greening in potato tubers, *Idaho Agric. Exp. Stn. Res. Bull.*, 16, 1, 1949.

Leberman, K. W., Nelson, A. I., and Steinberg, M. P., Postharvest changes of broccoli stored in modified atmospheres, *Food Technol.*, 22, 143, 1968.

Lewis, L. N., Coggins, C. W., Jr., Labanauskas, C. K., and Duggen, W. M., Jr., Biochemical changes associated with naturel and gibberellin A_3 delayed senescence in the "Navel" orange rind, *Plant Cell Physiol.*, 8, 151, 1967.

Li, P. H., Metabolism of pears in modified atmospheres, Ph.D. dissertation, Oregon State University, Corvallis, Oregon, 1963.

Licciardello, J. J., Nickerson, J. T. R., Proctor, B. E., and Campbell, C. L., Storage characteristics of some irradiated foods held at various temperatures above freezing. I. Studies with chicken meat and sweet potatoes, *Food Technol.*, 13, 398, 1959.

Lieberman, M., Asen, S., and Mapson, L. W., Ethylene oxide, an antagonist of ethylene in metabolism, *Nature*, 204, 756, 1964.

Lieberman, M. and Hardenburg, R. E., Effect of modified atmosphere on respiration and yellowing of broccoli at 75°F, *Proc. Am. Soc. Hortic. Sci.*, 63, 409, 1954.

Lieberman, M. and Kunishi, A. T., Thought on the role of ethylene in plant growth and development, in *Plant Growth Substances. 1970, 1972*, Carr, D. J., Ed., Springer-Verlag, New York, 1972.

Lieberman, M. and Mapson, L. W., Inhibition of the evolution of ethylene and the ripening of fruit by ethylene oxide, *Nature*, 196, 660, 1962.

Lipton, W. J., Market quality and rate of respiration of head lettuce held in low-oxygen atmospheres, USDA Marketing Res. Rep., 777, 1967.

Lipton, W. J., Controlled atmosphere effects on lettuce quality in simulated export shipments, USDA ARS 51-54, 1971.

Lipton, W. J. and Ceponis, M. J., Retardation of senescence and stimulation of oxygen consumption in head lettuce treated with N^6-benzyladenine, *Proc. Am. Soc. Hortic. Sci.*, 81, 379, 1962.

Lipton, W. J. and Matoba, F., Whitewashing to prevent sunburn of Crenshaw melons, *Hortic. Sci.*, 6, 343, 1971.

Littauer, F., Combined versus single treatments for the control of citrus fruit rots, *Ktavim Rec. Agric. Res. Stn.*, 6, 129, 1956.

Littlefield, N. A., Physiochemical and toxicological studies on controlled atmosphere storage of certain deciduous fruits, Ph.D. dissertation, Utah State University, Logan, Utah, 1968.

Liu, F. W., Storage of bananas in polyethylene bags with an ethylene absorbant, *Hortic. Sci.,* 5, 25, 1970.

Lodh, S. B., De, S., Mukherjee, S. K., and Bose, A. N., Storage of mandarin oranges. II. Effects of hormones and wax coatings, *J. Food Sci.,* 28, 519, 1963.

Lott, R. V. and Rice, R., Effects of preharvest sprays of 2,4,5-TP upon the ripening of "Jonathan," "Starking" and "Golden Delicious" apples, *Ill. Agric. Exp. Stn. Bull.,* No. 590, 1955.

Loucks, K. W. and Hopkins, E. F., A study of the occurrence of *Phomopsis* and of *Diplodia* rots in Florida oranges under various conditions and treatments, *Phytopathology,* 36, 750, 1946.

Lukton, A. and MacKinney, G., Effect of ionizing radiation on carotenoid stability, *Food Technol.,* 10, 630, 1956.

Lyons, J. M. and Pratt, H. K., An effect of ethylene on swelling of isolated mitochondria, *Arch. Biochem. Biophys.,* 104, 318, 1964.

MacLean, D. C. and Dewey, D. H., Reduction of decay of prepackaged apples with 2-aminobutane, *Mich. State Univ., Agric. Exp. Stn., Q. Bull.,* 47, 225, 1964.

Madsen, K. A., Salunkhe, D. K., and Simon, M., Morphological and biochemical changes in gamma-irradiated carrots (*Daucus carota* L.) and potatoes (*Solanum tuberosum* L.), *Radiat. Res.,* 10, 48, 1959.

Magie, R. O., Controlling gladiolus *Botrytis* bud rot with ozone gas, *Proc. Fla. State Hortic. Soc.,* 73, 373, 1961.

Magness, J. R. and Diehl, H. C., Physiological studies on apples in storage, *J. Agric. Res.,* 27, 33, 1924.

Mapson, L. W., Biosynthesis of ethylene and the ripening of fruit, *Endeavour,* 29, 29, 1970.

Markakis, P., Livingston, G. E., and Fagerson, I. S., Effects of cathode ray and gamma ray irradiation on the anthocyanin pigments of strawberries, *Food Res.,* 24, 520, 1959.

Marlott, R. B. and Stewart, J. K., Pink rib of head lettuce, *Plant Dis. Rep.,* 40, 742, 1956.

Marshall, R. E., Hammer, C. L., and Kremer, J. C., Retardation of ripening of fruit with the methyl ester of NAA, *Proc. Am. Soc. Hortic. Sci.,* 51, 95, 1948.

Mattus, G. E., Rate of respiration and volatile production of "Bartlett" pears following removal from air and controlled atmosphere storage, *Proc. Am. Soc. Hortic. Sci.,* 55, 199, 1950.

Mattus, G. E., Effects of succinic acid, 2,2-dimethylhydrazide (Alar) on apples, *Am. Soc. Hortic. Sci. 65th Ann. Marketing Prog. (Abstr. Insert),* 1968, 122.

Maxie, E. C. and Crane, J. C., 2,4,5-Trichlorophenoxyacetic acid: Effect on ethylene production by fruits and leaves of fig tree, *Science,* 155, 1548, 1967.

Maynard, J. A. and Swan, J. M., Organophosphorous compounds. I. 2-Chloroethanephosphonic acids as phosphorylating agents, *Aust. J. Chem.,* 16, 596, 1963.

McClure, T. T., Brown and *Rhizopus* rots of peaches as affected by hydrocooling, fungicides, and temperature, *Phytopathology,* 48, 322, 1958.

McCornack, A. A. and Hopkins, E. F., Decay control of Florida citrus with 2-aminobutane, *Proc. Fla. State Hortic. Soc.,* 77, 267, 1965.

McGill, J. N., Nelson, A. I., and Steinberg, M. P., Effects of modified storage atmosphere on ascorbic acid and other quality characteristics of spinach, *J. Food Sci.,* 31, 510, 1966.

McKee, R. K. and Boyd, A. E. W., Dry rot disease of the potato. IX. The effect of diphenyl vapor on dry rot infection of potato tubers, *Ann. Appl. Biol.,* 50, 89, 1962.

McKenzie, D. W., Influence of NAA on the preharvest drop of "Jonathan" apples — effects of repeated applications on preharvest drop and fruit quality, *N. Z. J. Sci. Technol.,* 35A, 45, 1953.

McKenzie, K. A., Respiration studies with lettuce, *Proc. Am. Soc. Hortic. Sci.,* 28, 244, 1931.

McMurchie, E. J., McGlasson, W. B., and Eaks, I. L., Treatment of fruit with propylene gives information about the biogenesis of ethylene, *Nature,* 237, 235, 1972.

Meredith, D. S., Studies on *Gloeosporium musarum* Cke. and Massee causing storage rots of Jamaican bananas. I. Anthracnose and its chemical control, *Ann. Appl. Biol.,* 48, 279, 1960.

Mikaelsen, K. and Roer, L., Improved storage ability of potatoes exposed to gamma radiation, *Acta Agric. Scand.,* 6, 145, 1956.

Miller, G. W., Brown, S. C., and Holmes, R. S., Chlorosis in soybeans are related to iron, phosphorous, bicarbonate and cytochrome oxidase activity, *Plant Physiol.,* 35, 619, 1960.

Miller, G. W. and Evans, H. J., Inhibition of plant cytochrome oxidase by bicarbonate, *Nature,* 178, 974, 1956.

Miller, G. W. and Hsu, W. P., Effects of carbon dioxide-bicarbonate mixture on oxidative phosphorylation by cauliflower mitochondria, *Biochem. J.,* 97, 615, 1965.

Monselise, S. P. and Goren, R., Changes in composition and enzymatic activity in flavedo of "Shamouti" oranges during the color break period as influenced by application of gibberellin and 2-chloroethyltrimethylammonium chloride, *Phyton,* 22, 61, 1965.

Morris, L. L. and Mann, L. K., Effect of a volatile from honeydew melons on the storage behavior of certain vegetables, *Proc. Am. Soc. Hortic. Sci.,* 47, 368, 1946.

Mothes, K., Engelbrecht, L., and Kulajewa, O., Uber die wirkung des kinetins auf stickstoff. Verteilung und eiweissynthese in isolierte blattern, *Flora* (Jena), 147, 445, 1959.

Murata, T., Ku, H. S., and Ogata, K., Studies on postharvest ripening and storage of banana fruits. III. Effects of growth regulating substances on the postharvest ripening of banana fruits, *J. Food Sci. Technol.,* 12, 461, 1965.

Nelson, K. E., Baker, G. A., and Gentry, J. P., Relation of decay and bleaching injury of table grapes to storage air velocity and relative humidity and to SO$_2$ treatment before and during storage, *Am. J. Enol. Vitic.,* 15, 93, 1964.

Nelson, K. E., Chiarappa, L., and Baker, G., Control of *Botrytis* decay in stored grapes with dibromotetrachloroethane, *Am. J. Enol. Vitic.,* 14, 105, 1963.

Novobranova, T. I., Mycoflora and infection of grapes with fungal rot during storage, *Vestn. SK-H Nauki* (Mosc.), 13(9), 74, 1970.

Oberbacher, M. F., The effect of Amchem 66-329 on degreening and respiration of "Valencia" oranges, *Hortic. Sci.,* 3(2), 122, 1968.

O'Brien, M., A system for mechanical harvesting and handling of pineapple, *Proc. 18th Int'l. Hortic. Cong.,* 1, 115, 1970.

Ogawa, J. M., Lyda, S. D., and Weber, D. J., 2,6-dichloro-4-nitroaniline effectiveness against *Rhizopus* fruit rot of sweet cherries, *Plant Dis. Rep.,* 45, 636, 1961.

Onslow, M. W., Oxidizing enzymes. III. The oxidizing enzymes of some common fruits, *Biochem. J.,* 14, 541, 1920.

Overcash, J. P., Some effects of certain growth-regulating substances on the ripening of "Concord" grapes, *Proc. Am. Soc. Hortic. Sci.,* 65, 54, 1955.

Padfield, C. A. S., The effect of preharvest anti-drop hormone spray upon "Granny Smith" apples in storage, *Orchardist N. Z.,* 22, 10, 1949.

Palmer, R. L., Lewis, L. N., Hield, H. Z., and Kumamoto, J., Abscission induced by betahydroxyethylhydrazine: Conversion of betahydroxyethylhydrazine to ethylene, *Nature,* 216, 1216, 1967.

Panalaskas, T. and Pellefier, O., The effect of storage on ascorbic acid content of gamma radiated potatoes, *Food Res.,* 25, 33, 1960.

Pantastico, E. B., Effect of gibberellic acid and kinetin on the growth respiration and chlorophyll content of rice and corn, Master's thesis, University of Philippines, Laguna, Philippines, 1964.

Pantastico, E. B. and Mendoza, D. B., Jr., Note: Production of ethylene and acetylene during respiration and charring, *Philipp. Agric.,* 53, 472, 1970.

Pantastico, E. B., Mendoza, D. B., Jr., Espino, V. C., Bondad, N. D., and Calara, E. R., Regulation of fruit ripening. I. Refrigerated controlled atmosphere, *Philipp. Agric.,* 54, 120, 1970.

Patil, B. C., Salunkhe, D. K., and Singh, B., Metabolism of solanine and chlorophyll in potato tubers as affected by light and specific chemicals, *J. Food Sci.,* 36, 474, 1971a.

Patil, B. C., Singh, B., and Salunkhe, D. K., Formation of chlorophyll and solanine in Irish potato (*Solanum tuberosum* L.) tubers and their control by gamma radiation and CO$_2$ enriched packaging, *Lebensm.-Wiss. U. Technol.,* 4, 123, 1971b.

Patterson, D. R. and Wittwer, S. H., Further investigations on the use of maleic hydrazide as a sprout inhibitor for onions, *Proc. Am. Soc. Hortic. Sci.,* 62, 409, 1953.

Pederson, S., The effect of ionizing radiations on sprout prevention and chemical composition of potatoes, *Food Technol.,* 10, 532, 1956.

Pennock, W. and Maldonaldo, G., Hot-water treatment of mango fruits to reduce anthracnose decay, *J. Agric. Univ. P. R.,* 46, 272, 1962.

Pentzer, W. T., Methods of applying sodium bisulfite to grape packages for mold control, *Blue Anchor,* 16(7), 2, 1939.

Pierson, C. F., Fungicides for reduction of post-harvest decay of sweet cherries, *Proc. Wash. State Hortic. Assoc.,* 54, 115, 1958.

Pierson, C. F., Postharvest fungicide treatments for reduction of decay in Anjou pears, *Plant Dis. Rep.,* 44, 64, 1960.

Porter, L. K. and Thorne, D. W., Interrelations of carbon dioxide and bicarbonate ions in causing plant chlorosis, *Soil Sci.,* 79, 373, 1955.

Post, F. J., Coblentz, W. S., Chou, T. W., and Salunkhe, D. K., Influence of phosphate compounds on certain fungi and their preservative effects on fresh cherry fruit (*Prunus cerasus,* L.), *Appl. Microbiol.,* 16, 138, 1968.

Pridham, J. B., Determination of sugars on paper chromatograms with p-anisidine hydrochloride, *Anal. Chem.,* 28, 1967, 1956.

Pryor, D. E., Reduction of post-harvest spoilage in fresh fruits and vegetables destined for long distance shipment, *Food Technol.,* 4, 57, 1950.

Rabinowitch, H. D., Rudich, J., and Kedar, N., The effect of Ethrel on ripening of tomato and melon fruits, *Isr. J. Agric. Res.,* 20(1), 47, 1970.

Rahman, A. R., Schafer, G., Dinicola, T. J., and Westcott, D. E., Shelf life of tomatoes as affected by a postharvest treatment and storage conditions, U.S. Natl. Tech. Inform. Serv., AD Rep. No. 746254, 1972.

Rahman, A. R., Schafer, G., and Westcott, D. E., Storage life of lettuce as affected by controlled atmosphere system, U.S. Army Tech. Report 70-48-FL, 1969.

Rakitin, Y. V., Krylov, A. V., and Tsrakanova, G. A., Metabolic transformation of methyl alcohol introduced into fruits to hasten ripening, *Dokl. Bot. Sci.* (English Transl.), 116, 874, 1957.

Ramsey, G. B. and Butler, L. F., Injury to onions and fruits caused by exposure to ammonia, *J. Agric. Res.,* 37, 339, 1928.

Ramsey, G. B., Smith, M. A., and Heiberg, B. C., Fungistatic action of diphenyl on citrus fruit pathogens, *Bot. Gaz.,* 106, 74, 1944.

Ranson, S. L., Zymasis and acid metabolism in higher plants, *Nature,* 172, 252, 1953.

Ranson, S. L., Walker, D. A., and Clarke, I. D., The inhibition of succinic oxidase by high CO$_2$ concentrations, *Biochem. J.,* 66, 57, 1957.

Rao, S. N. and Wittwer, S. H., Further investigation on the use of maleic hydrazide as a sprout inhibitor for potatoes, *Am. Potato J.,* 32, 51, 1955.

Rasmussen, G. K. and Cooper, W. C., Abscission of citrus fruits induced by ethylene-producing chemicals, *Proc. Am. Soc. Hortic. Sci.,* 93, 191, 1968.

Reeve, R. M., Histological and histochemical changes in developing and ripening peaches. I. The catechol tannins, *Am. J. Bot.,* 46(3), 210, 1959.

Richmond, A. E. and Lang, A., Effect of kinetin on protein content and survival of detached Xanthium leaves, *Science,* 125, 650, 1957.

Robinson, R. W., Wilcynski, H., Dennis, F. G., Jr., and Bryan, H. H., Chemical promotion of tomato fruit ripening, *Proc. Am. Soc. Hortic. Sci.,* 93, 823, 1968.

Rodrigues, J. and Subramanyam, H., Effect of preharvest spray of plant growth regulators on size, composition and storage behaviour of "Coorg" mandarins (*Citrus reticulata* Blanco), *J. Sci. Food Agric.,* 17, 425, 1966.

Roistacher, C. N., Eaks, I. L., and Klotz, L. J., Ammonia gas to control blue-green mold decay of citrus fruits, *Plant Dis. Rep.,* 39, 202, 1955.

Roistacher, C. N., Klotz, L. J., and Eaks, I. L., Blue-green mold on citrus fruits, *Calif. Agric.,* 11, 11, 1957.

Roistacher, C. N., Klotz, L. J., and Garber, M. J., Tests with volatile fungicides in packages on citrus fruits during shipment to eastern markets, *Phytopathology,* 50, 855, 1960.

Russell, A., The natural tannins, *Chem. Rev.,* 17, 155, 1935.

Russo, L., Jr., Dostal, H. C., and Leopold, A. C., Chemical regulations of fruit ripening, *Bioscience,* 18, 108, 1968.

Rygg, G. L., Wilson, C. W., and Garber, M. J., Effect of biphenyl treatment and carton ventilation on decay and spoilage of California lemons in overseas shipments, USDA, Marketing Res. Rep., 500, 1, 1961.

Ryugo, K. and Sachs, R. M., *In vitro* and *in vivo* studies of Alar (1,1-dimethylaminosuccinamic acid, B-995) and related substances, *J. Am. Soc. Hortic. Sci.,* 94, 529, 1969.

Sacher, J. A., Studies of permeability, RNA and protein turnover during ageing of fruit and leaf tissues, *Symp. Soc. Exp. Biol.,* 21, 269, 1967.

Saha, A. K., Effect of postharvest treatment with growth regulators on the ripening and chemical composition of guava (*Psidium guajava* L.) fruits, *Indian J. Hortic.,* 28(1), 11, 1971.

Salas, R. P., Ripening of "Dungulan" and "Saba" bananas treated with acetylene, Undergraduate thesis, University of Philippines, Laguna, 1957.

Salem, E. A., Fahmy, B. A., Ibrahim, R., and Salama, S. B., Control of storage decay in oranges, *Agric. Res. Rev.,* 48(5), 104, 1970.

Salunkhe, D. K., Histological and histochemical changes in gamma-irradiated lima beans (*Phaseolus lunatus* L.), *Nature,* 179, 585, 1957.

Salunkhe, D. K., Physiological and bio-chemical effects of gamma radiation on tubers of Jerusalem artichoke, *Bot. Gaz.,* 120, 180, 1959.

Salunkhe, D. K., Anderson, J. L., and Patil, J. B., Hastening green wrap tomato ripening—with Ethrel, *Utah Sci.,* 32, 6, 1971a.

Salunkhe, D. K., Cooper, G. M., Dhaliwal, A. S., Boe, A. A., and Rivers, A. L., On storage of fruits: Effects of pre- and postharvest treatments, *Food Technol.,* 16, 119, 1962.

Salunkhe, D. K., Gerber, R. K., and Pollard, L. H., Physiological and chemical effects of gamma radiation on certain fruits and vegetables and their products, *Proc. Am. Soc. Hortic. Sci.,* 74, 423, 1959a.

Salunkhe, D. K. and Norton, R. A., Packaging treatments extend storage life of fruit, *Utah Agric. Exp. Stn. Farm Home Econ. Sci.,* 21, 18, 1960.

Salunkhe, D. K., Pollard, L. H., and Gerber, R. K., Effects of gamma radiation dose, rate, and temperature on the taste preference and storage life of certain fruits, vegetables, and their products, *Proc. Am. Soc. Hortic. Sci.,* 74, 414, 1959b.

Salunkhe, D. K., Pollard, L. H., Gerber, R. K., Wilcox, E. B., and Simon, M., Packaging effects on the flavor and shelf-life of gamma radiated fresh fruits and vegetables, *Packaging Eng.,* 4, 39, 1959c.

Salunkhe, D. K. and Simon, M., Further studies on effects of gamma radiation on fruits and vegetables, *Food Technol.,* 14(4), 28, 1960.

Salunkhe, D. K., Wittwer, S. H., Wheeler, E. J., and Dexter, S. T., The influence of a preharvest foliar spray of maleic hydrazide on specific gravity of potatoes and quality of potato chips, *Food Res.,* 18, 191, 1953.

Salunkhe, D. K. and Wu, M. T., Effects of low oxygen atmosphere on the ripening behavior and certain biochemical changes of tomatoes, *J. Am. Soc. Hortic. Sci.,* 98, 12, 1973a.

Salunkhe, D. K. and Wu, M. T., Effects of subatmospheric pressure storage on the ripening behavior and some chemical changes on certain deciduous fruits, *J. Am. Soc. Hortic. Sci.,* 98, 113, 1973b.

Salunkhe, D. K., Wu, M., Wu, M. T., and Singh, B., Effects of Telone and Nemagon on essential nutritive components and the respiratory rates of carrot (*Daucus carota* L.) roots and sweet corn (*Zea mays* L.) seeds, *J. Am. Soc. Hortic. Sci.,* 96, 357, 1971b.

Sawyer, R. L. and Dallyn, S. L., Vaporized chemical inhibitors and irradiation, two new methods of sprout control for tubers and bulb crops, *Proc. Am. Soc. Hortic. Sci.,* 67, 514, 1956.

Sawyer, R. L. and Dallyn, S. L., Vaporized chemicals control sprouting in stored potatoes, Farm. Res., July, 1957, 6.

Schomer, H. A. and McCollock, L. P., Ozone in relation to storage of apples, USDA Circ. 765, 1-24, 1948.

Schreiber, J. S. and Highlands, M. E., A report on the commercial storage of irradiated white potatoes, *Maine Farm Res.,* 5(4), 19, 1958.

Scott, K. J., McGlasson, B., and Roberts, E. A., Ethylene absorbent increases storage life of bananas packed in polyethylene absorbent, *Agric. Gaz. N. S. W.,* 79, 52, 1968.

Scott, K. J., McGlasson, B., and Roberts, E. A., Potassium permanganate as an ethylene absorbent in polyethylene bags to delay ripening of bananas during storage, *Aust. J. Exp. Agric. Anim. Husb.,* 10, 237, 1970.

Sfakiotakis, E. M. and Dilley, D. R., Induction of autocatalytic ethylene production in apple fruits by propylene in relation to maturity and oxygen, *J. Am. Soc. Hortic. Sci.,* 98, 504, 1973.

Shah, A. K. and Ghosh, S. K., Some studies on postharvest physiology of banana (*Musa paradisiaca* var. "Kabuli") fruits as influenced by maturity and growth regulators, Proc. 3rd Int'l. Hortic. Symp. Trop. Subtrop. Hortic., 1972, 103.

Sharma, J. N., Art of preventing decay, U.S. Patent 2,054,392, 1936.

Simmonds, J. H., Dipping of winter bananas, *Queensl. Agric. J.,* 68, 274, 1949.

Sims, W. L., Effects of Ethrel on fruit ripening of tomatoes . . . greenhouse, field and postharvest trials, *Calif. Agric.,* 23, 12, 1969.

Sims, W. L., Collins, H. B., and Gledhill, B. L., Ethrel effects on fruit ripening of peppers, *Calif. Agric.,* 24(2), 4, 1970.

Singh, B., Littlefield, N. A., and Salunkhe, D. K., Effect of CA storage on amino acids, organic acids, sugar, and rate of respiration of "Lambert" sweet cherry fruit, *J. Am. Soc. Hortic. Sci.,* 95, 458, 1970.

Singh, B., Wang, D. J., Salunkhe, D. K., and Rahman, A. R., Controlled atmosphere storage of lettuce. II. Effects on biochemical composition of the leaves, *J. Food Sci.,* 37, 52, 1972a.

Singh, B., Yang, C. C., Salunkhe, D. K., and Rahman, A. R., Controlled atmosphere storage of lettuce. I. Effects on quality and rate of respiration of lettuce heads, *J. Food Sci.,* 37, 48, 1972b.

Singh, N. S., Krishnaprakash, M. S., Moorthy, N. V. N., Narasimham, P., and Nair, K. G., Effect of treatment with hot water on ripening and lycopene content of tomatoes, *J. Food Sci. Technol.,* 6, 18, 1969.

Smith, W. L., Jr., Streptomycin sulfate for the reduction of bacterial soft rot of packaged spinach, *Phytopathology,* 45, 88-90, 1955.

Smith, W. L., Jr., Reduction of postharvest brown rot and *Rhizopus* decay of eastern peaches with hot water, *Plant Dis. Rep.,* 46, 861, 1962.

Smith, W. L., Jr. and Bassett, R. D., Reduction of postharvest decay of peaches and nectarines with heat treatments. I. Factors involved in effective heat treatment of peaches, USDA Marketing Res. Rep., 643, 5, 1964.

Smock, R. M., The storage of apples, *Cornell Exp. Bull.,* No. 440, 1958a.

Smock, R. M., Controlled atmosphere storage of apples, *Cornell Exp. Bull.,* No. 759, 1958b.

Smock, R. M., Methods of storing bananas, *Philipp. Agric.,* 51, 501, 1967.

Smock, R. M. and Allen, F. W., Soluble pectin changes in gas-stored fruit, *Proc. Am. Soc. Hortic. Sci.,* 35, 184, 1937.

Smock, R. M. and Blanpied, G. D., Effect of modified technique in CA storage of apples, *Proc. Am. Soc. Hortic. Sci.,* 87, 73, 1966.

Smock, R. M., Edgerton, L. J., and Hoffman, M. B., Some effects of stop-drop auxins and respiratory inhibitors on the maturity of apples, *Proc. Am. Soc. Hortic. Sci.,* 63, 211, 1954.

Smock, R. M., Martin, D., and Padfield, D. H. S., Effect of N^6-benzyladenine on the respiration and keeping quality of apples, *Proc. Am. Soc. Hortic. Sci.,* 81, 51, 1962.

Smock, R. M. and Yatsu, L., Removal of carbon dioxide from controlled atmosphere storage with water, *Proc. Am. Soc. Hortic. Sci.,* 75, 53, 1960.

Smoot, J. J. and Melvin, C. F., Hot water as a control for decay of oranges, *Proc. Fla. State Hortic. Sci.,* 76, 322, 1964.

Smoot, J. J. and Segall, R. H., Hot water as a postharvest control of mango anthracnose, *Plant Dis. Rep.,* 47, 739, 1963.

Soni, S. L., Chauhan, K. S., and Jain, S. C., Effect of plant growth regulators, wax emulsions and their combinations on the storage behavior and physiochemical changes during storage of banana (*Musa paradisiaca* Linn.), Proc. 3rd Int. Symp. Trop. Subtrop., 1972, 77.

Southwick, F. W., The influence of nitrogen level on the rate of ripening and color development of apples sprayed with growth regulating substances, *Proc. Am. Soc. Hortic. Sci.,* 63, 225, 1953.

Southwick, F. W. and Lachman, W. H., The effect of maleic hydrazide and water on the rate of respiration of harvested tomato fruits, *Proc. Am. Soc. Hortic. Sci.,* 61, 388, 1953.

Spalding, D. H., Postharvest use of benomyl and thiabendazole to control blue mold rot development in pears, *Plant Dis. Rep.,* 54, 655, 1970.

Sparrow, A. H. and Christensen, E., Improved storage quality of potato tubers after exposure to Co^{60} gammas, *Nucleonics,* 12(8), 16, 1954.

Stewart, J. K., Hot water and fungicides for control of mold on cantaloups, USDA Marketing Res. Rep., 986, 1, 1973.

Stewart, J. K., Ceponis, M. J., and Beraha, L., Modified atmosphere effects on the market quality of lettuce shipped by rail, USDA Marketing Res. Rep. No. 863, 1973, 10.

Stewart, J. K., Harvey, J. M., Ceponis, M. J., and Wright, W. R., Nitrogen – its effects on transit temperatures and market quality of western lettuce shipped in piggyback trailers, USDA Marketing Res. Rep. No. 832, 1966.

Stewart, J. K. and Uota, M., Carbon dioxide injury and market quality of lettuce held in controlled atmospheres, *J. Am. Soc. Hortic. Sci.*, 96, 27, 1971.

Stewart, J. K. and Uota, M., Carbon dioxide injury to lettuce as influenced by carbon monoxide and oxygen levels, *Hortic. Sci.*, 189, 1972.

Stewart, J. K. and Wells, J. M., Heat and fungicide treatment to control decay of cantaloups, *J. Am. Soc. Hortic. Sci.*, 95, 226, 1970.

Stewart, W. S., Effects of 2,4-dichlorophenoxyacetic acid and 2,4,5-trichlorophenoxyacetic acid on citrus fruit storage, *Proc. Am. Soc. Hortic. Sci.*, 54, 109, 1949.

Stewart, W. S., Maturity and ripening as influenced by application of plant regulators, *Plant Regulators in Agriculture*, Turkey, H. B., Ed., J. Wiley and Sons, Inc., New York, 1956, 132.

Stewart, W. S., Palmer, J. R., and Hield, H. Z., Packinghouse experiments on the use of 2,4-D and 2,4,5-T to increase storage life of lemons, *Proc. Am. Soc. Hortic. Sci.*, 59, 327, 1952.

Subramanyam, H., Moorthy, N. V. N., Subhadra, N. V., and Mathu, M., Control of spoilage and inhibition of ripening in "Alphonso" mangoes by fumigation, *Trop. Sci.*, 11(2), 120, 1969.

Subramanyam, H. and Sebastian, K., Effect of succinic acid, 2,2-dimethyl hydrazide on carotene development in "Alphonso"mango, *Hortic. Sci.*, 5, 160, 1970.

Sugiura, M., Umenura, K., and Oota, Y., The effect of kinetin on protein level of tobacco leaf disks, *Physiol. Plant*, 15, 457, 1962.

Swain, T., Economic importance of flavonoid compounds, *The Chemistry of Flavonoid Compounds*, Geissman, T. A., Ed., MacMillan Co., New York, 1962, 513.

Szkolnik, M. and Hamilton, J. M., Control of peach leaf curl with Omadine and of brown rot with Omadine and certain antibiotics, *Plant Dis. Rep.*, 41, 289, 1957.

Tabing, L. M. and Gonzales, L. G., Influence of different applications of maleic hydrazide in the keeping quality of onion, *Philipp. Agric.*, 40, 627, 1956.

Teaotia, S. S., Tripathi, C. S., and Singh, R. M., Effect of growth substances on ripening and quality of guava (*Psidium guajava* L.), *J. Food Sci. Technol.*, 9, 38, 1972.

Tehrani, G. and Tibor, F., The effect of succinic 2, 2-dimethylhydrazide (Alar) on color development of "Montmorency" tart cherry, *Hortic. Sci.*, 4(2), 151, 1969.

Teixeira, A. R. and Baptista, J. E., Preliminary study on irradiation of agricultural product, *Agron. Lusit.*, 30, 101, 1968.

Teskey, B. J. E. and Francis, F. J., Color changes in skin and flesh of stored "McIntosh" apple sprayed with 2,4, 5-trichlorophenoxypropionic acid, *Proc. Am. Soc. Hortic. Sci.*, 63, 220, 1954.

Thimann, K. V. and Manmohan, K. L., Changes in nitrogen in pea stem sections under the action of kinetin, *Physiol. Plant*, 13, 165, 1960.

Thomas, M., The controlling influence of CO_2, *Biochem. J.*, 19, 927, 1925.

Thompson, A. H., Further experiments with 2,4,5-trichlorophenoxypropionic acid sprays for control of the preharvest drop of apples, *Proc. Am. Soc. Hortic. Sci.*, 60, 175, 1952.

Thornton, N. C., Carbon dioxide storage of fruits and vegetables and flowers, 2nd. Eng. Chem., 22, 1186, 1930.

Thornton, N. C., Carbon dioxide storage. XI. The effect of CO_2 on the ascorbic acid (vitamin C) content of some fruits and vegetables, *Proc. Am. Soc. Hortic. Sci.*, 35, 200, 1937.

Timm, H., Bishop, J. C., and Hoyle, B. J., Investigations with maleic hydrazide on potatoes. I. Effect of time of application and concentration upon potato performance, *Am. Potato J.*, 36(4), 115, 1959.

Tolle, W. E., Hypobaric Storage of mature-green tomatoes, USDA Marketing Res. Rep., 842, 1969.

Tomkins, R. G., Wraps for the prevention of rotting of fruit, *G.B. Dept. Sci. 2nd Research, Food Invest. Board Rep.*, 129-131, 1936.

Trout, S. A., Experiments on the storage of pears in artificial atmospheres, *J. Pomol. Hortic. Sci.*, 8, 78, 1930.

Tsai, C. C. and Chiang, M. N., The effect of ethylene-producing chemicals on the degreening rate of detached lemons and "Satsuma" oranges (*Citrus unshiu* Marc.), *Mem. Coll. Agric. Nat'l Taiwan Univ.*, 11(1), 14, 1970.

Tuli, V., Dedolph, R. R., and Wittwer, S. H., 1962. Effects of N^6-benzyladenine and dehydroacetic acid on the storage behavior of cherries and strawberries, *Mich. Agric. Exp. STn. Q. Bull.*, 45(2), 223, 1962.

Tyler, K., May, D., and Miller, K., Ethrel sprays reduce number of pickings in hand-harvested cantaloups, *Calif. Agric.*, 24(4), 6, 1970.

Vendrell, M., Reversion of senescence: Effects of 2,4-dichlorophenoxyacetic acid on respiration, ethylene production, and ripening of banana fruit slices, *Aust. J. Biol. Sci.*, 22, 601, 1969.

Vendrell, M., Relationship between internal distribution of exogenous auxins and accelerated ripening of banana fruit, *Aust. J. Biol. Sci.*, 23, 1133, 1970.

Walker, J. R. L., Studies on the enzymic browning of apples. II. Properties of apple polyphenol oxidase, *Aust. J. Biol. Sci.*, 17, 360, 1964.

Wang, D. J., Certain changes in chemical composition of lettuce stored in controlled atmosphere, Masters thesis, Utah State University, Logan, Utah, 1971.

Wang, S. S., Haard, N. F., and DiMarco, G. R., Chlorophyll degradation during controlled atmosphere storage of asparagus, *J. Food Sci.*, 36, 657, 1971.

Wankier, B. N., Effects of controlled atmosphere storage on specific biochemical changes in apricot and peach fruits, Ph.D. dissertation, Utah State University, Logan, Utah, 1970.

Wankier, B. N., Salunkhe, D. K., and Campbell, W. F., Effects of CA storage on biochemical changes in apricot and peach fruit, *J. Am. Soc. Hortic. Sci.,* 95, 604, 1970.

Watada, A. E., Morris, L. L., and Rappaport, L., Modified atmosphere effects of lettuce, Fruit and Vegetable Perishable Handling Conference Proc., University of Calif., Davis, Calif., 1964.

Watada, A. E. and Scott, E. G., Metabolic reduction of harvested strawberry fruits treated with sodium dehydroacetate, *Proc. 18th Int'l. Hortic. Cong.,* 1, 8, 1970.

Weinberger, J. H., Effect of 2,4,5-trichlorophenoxyacetic acid on ripening of peaches in Georgia, *Proc. Am. Soc. Hortic. Sci.,* 51, 115, 1951.

Wiant, L. S. and Bratley, C. O., Spoilage of fresh fruits and vegetables in rail shipments unloaded at New York City, 1935-42, USDA Circular No. 773, 1, 1948.

Williams, M. W., Physical and biochemical studies on core breakdown on "Bartlett" pears stored in controlled atmosphere. Ph.D. dissertation, Washington State University, Pullman, Wash., 1961.

Winston, J. R., Decay of Florida citrus fruits and its control, *Citrus Ind.,* 29, 5, 1948.

Wishnetsky, T., Livingston, G. E., Francis, F. J., and Fagerson, I. S., Effects of gamma ray irradiation on color and chlorophyll retention in green beans and broccoli, *Food Technol.,* 13, 352, 1959.

Wittwer, S. H., Dedolph, R. R., Tuli, V., and Gilbart, D., Respiration and storage deterioration in celery (*Apium graveolens* L.) as affected by postharvest treatments with N^6-benzylaminopurine, *Proc. Am. Soc. Hortic. Sci.,* 80, 408, 1962.

Wittwer, S. H. and Dedolph, R. R., Some effects of kinetin on the growth and flowering of intact green plants, *Am. J. Bot.,* 50(4), 330, 1963.

Wittwer, S. H. and Hansen, C. M., The reduction of storage losses in sugar beets by preharvest foliage sprays of maleic hydrazide, *Agron. J.,* 43, 340, 1951.

Wittwer, S. H. and Paterson, D. R., Inhibition of sprouting and reduction of storage losses in onions, potatoes, sugar beets, and vegetable root crops by spraying plants in the field with maleic hydrazide, *Mich. Agric. Exp. Stn. Q. Bull.,* 34, 3, 1951.

Wittwer, S. H. and Sharma, R. C., The control of storage sprouting in onions of preharvest foliage sprays of maleic hydrazide, *Science,* 112, 597, 1950.

Wittwer, S. H., Sharma, R. C., Weller, L. E., and Sell, H. M., The effect of preharvest foliage sprays of certain growth regulators on sprout inhibition and storage quality of carrots and onions, *Plant Physiol.,* 25, 529, 1950.

Workman, M., Patterson, M. E., Ellis, N. K., and Heiligman, F., The utilization of ionizing radiation to increase the storage life of white potatoes, *Food Technol.,* 14, 395, 1960.

Workman, M., Pratt, H. K., and Morris, L. L., Studies on the physiology of tomato fruits. I. Respiration and ripening behavior at 20 degrees C as related to date of harvest, *Proc. Am. Soc. Hortic. Sci.,* 69, 352, 1957.

Wu, M., Singh, B., Wu, M. T., Salunkhe, D. K., and Dull, G. G., Effects of certain soil fumigants on essential nutritive components and the respiratory rates of carrot roots, *Hortic. Sci.,* 5, 221, 1970.

Wu, M. T., Jadhav, S. J., and Salunkhe, D. K., Effects of subatmospheric pressure storage on ripening of tomato fruits, *J. Food Sci.,* 37, 952, 1972.

Wu, M. T. and Salunkhe, D. K., Control of chlorophyll and solanine formation in potato tubers by oil and diluted oil treatment, *Hortic. Sci.,* 7, 466, 1972a.

Wu, M. T. and Salunkhe, D. K., Control of chlorophyll and solanine synthesis and sprouting of potato tubers by hot paraffin wax, *J. Food Sci.,* 37, 629, 1972b.

Wu, M. T. and Salunkhe, D. K., Fungistatic effects of subatmospheric pressures, *Experientia,* 28, 866, 1972c.

Wu, M. T. and Salunkhe, D. K., Inhibition of chlorophyll and solanine formation and sprouting of potato tubers by oil dipping, *J. Am. Soc. Hortic. Sci.,* 97, 614, 1972d.

Wu, M. T., Singh, B., Theurer, J. C., Olson, L. E., and Salunkhe, D. K., Control of sugar loss in sugar beet during storage by chemicals and modified atmosphere and certain associated physiological changes, *J. Am. Soc. Sugar Beet Technol.,* 16, 117, 1970.

Yang, C. C., Effect of controlled atmosphere storage on quality and certain physiological characteristics of lettuce, Master's thesis, Utah State University, Logan, Utah, 1971.

Young, R. E., Jahn, O., Cooper, W. C., and Smoot, J. J., Preharvest sprays with 2-chloroethylphosphonic acid to degreen "Robinson" and "Lee" tangerine fruits, *Hortic. Sci.,* 5, 268, 1970.

Young, R. E., Romani R. J., and Biale, J., Carbon dioxide effects on fruit respiration. II. Response of avocados, bananas, and lemons, *Plant Physiol.,* 37, 416, 1962.

Zienlinski, O. B., Effects of 2,4,5-trichlorophenoxyacetic acid on the maturation of prunes, *Proc. Am. Soc. Hortic. Sci.,* 58, 65, 1951.

Zink, F. W., N^6-benzyladenine, a senescence inhibitor for green vegetables, *J. Agric. Food Chem.,* 9, 304, 1961.

INDEX

Through-flow drying, 46
Thyrotoxin, 22
Toxicological, 117
Trace elements, 65
Transaminase, 100
s-Triazines, 32, 87, 88
  of forage crops, 88
  preharvesting applications, 32
  protein contents, 88
2,4,5-Trichlorophenoxyacetic acid (2,4,5-T), 129
2,4,5-Trichlorophenoxypropionic acid (2,4,5-TP), 129
Triflurolin, 85
2,3,5-Triiodobenzoic acid (TIBA), 82
Trimming, 24
Tunnel dehydrators, 40
Tunnel drying, 44
"Tunta", 39
Turnip (*Brassica rapa L.*), 1

U

Ultrastructure of the cotyledon parenchyma cells, 104
Undernutrition, 79
Utilization, 70

V

Vacuum puff drying, 50
Vegetables, cruciferous, 4, 6, 7, 13, 32
  fats and carbohydrates, 6
  highly-nutritious, 32
  linoleic and linolenic, 7
  low-calorie foods, 32

nutritive contribution, 4
per capita consumption, 6
production statistics, 4
proteins, 7
quality, 7
quantity, 7
sources of minerals, 13
utilization, 7
Vitamins, 16, 81
  cruciferous vegetables, 16
    folic acid, 16
    niacin, 16
    pantothenic acid, 16
    pyridoxine, 16
    riboflavin, 16
    thiamine, 16
    vitamin A (carotene), 16
    vitamin C, 16
  fat-soluble, 81
  water soluble, 81

W

Washing, 24, 155
Water activity, 65
Watercress (*Nasturtium officinale* R. Br.), 1
Water quality, 25, 26
  nutritive value, 26
  wholesomeness of processed products, 26

Y

Yield, 80